LASER
MODELING

A Numerical Approach with Algebra and Calculus

LASER MODELING

A Numerical Approach with Algebra and Calculus

Mark Steven Csele

CRC Press
Taylor & Francis Group
Boca Raton   London   New York

CRC Press is an imprint of the
Taylor & Francis Group, an **informa** business

CRC Press
Taylor & Francis Group
6000 Broken Sound Parkway NW, Suite 300
Boca Raton, FL 33487-2742

First issued in paperback 2017

© 2014 by Taylor & Francis Group, LLC
CRC Press is an imprint of Taylor & Francis Group, an Informa business

No claim to original U.S. Government works

Version Date: 20140210

ISBN 13: 978-1-4665-8250-7 (hbk)
ISBN 13: 978-1-138-07199-5 (pbk)

Library of Congress Cataloging-in-Publication Data

Csele, Mark.
    Laser modeling : a numerical approach with algebra and calculus / author, Mark Steven Csele.
        pages cm
    Includes bibliographical references and index.
    ISBN 978-1-4665-8250-7 (hardback)
    1. Lasers--Mathematical models. 2. Laser beams--Mathematical models. 3. Numerical analysis. I. Title.

QC688.C75 2014
621.36'60151--dc23   2013049550

Visit the Taylor & Francis Web site at
http://www.taylorandfrancis.com

and the CRC Press Web site at
http://www.crcpress.com

Contents

Preface

The approach taken in this book is simple: present laser theory in an understandable way and one that can be applied immediately, and numerically, to real laser systems. With that in mind, the approach in this text is to present each theory along with a real, solved example—in most cases, based on commercial lasers. As a professor of laser science, I am fortunate to have a lab equipped with many different types of lasers; many of those lasers are included here in examples.

In making the theory "accessible," both a calculus-based and an algebraic approach are shown in tandem; a prime example of this is the presentation of both the calculus-based Rigrod model and an algebra-based model for the prediction of various laser parameters in Chapters 3 and 4. Readers drawn to numerically grounded solutions to problems (dare we say "engineers"?) will find the algebraic approach a refreshing demonstration of how concepts actually work and are applied, while those with more mathematical thought processes will appreciate the complementary calculus-based models. Either way, the results are similar (and, as I tell my students, it doesn't matter how you learn it). As an educator, I appreciate the fact that we all learn in different ways.

The actual use of algebra-based solutions originated with our four-year bachelor program at Niagara College. Although at the inception of the program we intended to use calculus-based theory exclusively, it became apparent that students were spending more time on the math than on the concepts (i.e., they were often "buried in the math," with the mathematical rigor of solutions getting in the way of understanding the concepts). In an attempt to reverse that situation, an algebra-based solution was presented in conjunction with the traditional approach. It was very well received by students and has been a staple of my laser courses since.

An example of the value of the algebraic approach is the Q-switch model presented in Chapter 6. This model predicts how inversion builds in time, and it was found that students spent more time developing and solving derivatives than understanding the concepts involved; by using an algebraic model, one could easily see how inversion builds, then decays, in a laser medium. This nicely demonstrates the concept, after which implementing a more complex calculus-based model is considerably easier for many students.

Although some knowledge of basic physics (such as the nature of light, emission of radiation, and some basic atomic physics) is assumed, the text begins with Chapter 1 providing a review of basic concepts and definitions of terms. A wide variety of approaches can be found in various texts and research papers, so it is worthwhile to provide a frame of reference. Most material in this chapter is a basis for concepts explored later in the text.

Since all models start with, or require, an accurate gain threshold equation, Chapter 2 is devoted entirely to this topic. While superficially simple, the formulation of this equation is more complex to accurately reflect the laser system and all components—one size does NOT fit all here, and the basic threshold equation familiar to many readers does not apply to all lasers. For ultimate accuracy, a unique threshold equation is required for each laser configuration. Most importantly, the methodology of developing this equation is outlined and many examples are given such that readers can formulate an equation specific for their application. Once the equation is developed, several simple applications of this equation are outlined, including the determination of small-signal gain (one of the key parameters of all lasing media).

The second most important concept, gain saturation, is covered in Chapter 3 along with one of the most powerful models, the pass-by-pass model. Although this model could be implemented in any computer programming language, it is shown here completed on a spreadsheet, since this is essentially a universal tool on all computers. From a learning perspective, this model is an excellent vehicle to demonstrate a host of concepts. This algebraic model may be used with almost any configuration of laser and can predict a host of performance parameters. It allows rapid implementation of "what if?" scenarios. Spreadsheet-based models used in this text are available from the publisher's website as well so that readers may modify them to suit their application.

Armed with the basic concepts from Chapters 2 and 3, the calculus-based Rigrod approach is outlined in Chapter 4. This approach, the "gold standard" for laser work, is developed in a simplified manner and applied to several lasers. Limitations of the approach are outlined and comparison made to the algebraic approach of Chapter 3. Used in tandem, both approaches (algebraic and Rigrod) allow maximum understanding of laser processes (appreciating, again, that we all learn in different ways).

With the increase in the number of solid-state lasers available, Chapter 5 outlines thermal effects on these (as well as other) lasers. Specifically, thermal population of the lower-lasing levels of many new quasi-three-level materials, as well as the effect of temperature on diode and DPSS lasers, is examined in detail. This topic is one that seems to be lacking in many texts (perhaps for no other reason than the relative increase in importance of solid-state lasers in the past few years and the ongoing search for newer and more efficient materials—what was a "lab curiosity" only a few years ago is "commercially viable" today).

Chapter 5 also presents an outline of how the convolution technique can be applied to predict the effect of temperature drift on a DPSS system. The effects of temperature on a pump diode, coupled with the characteristics of solid-state amplifier materials, lead to systems that are often unexpectedly sensitive to temperature change. Convolution, long used in the world of digital signal processing (DSP) for filtering signals, is implemented here (again using a spreadsheet for simplicity) to predict such effects.

Q-switching, an important technique integral to many lasers, is presented in Chapter 6. Having outlined the technology involved (i.e., how real Q-switches actually work), a simple model is presented to predict the output power of a Q-switched laser. Numerical examples are given outlining application to both a "normal" Q-switched laser as well as a considerably more unusual double-pulsed laser system (which serves as an excellent example of application of the theory).

Chapter 7 serves as a bit of a "crash course" in non-linear optics, which are so prevalent in modern laser systems. The theory of non-linear radiation generation is presented with practical examples given on determination of irradiance for a Q-switched laser to determine the suitability of a particular non-linear material. A simple model for predicting the optimal crystal length is developed ("simple," since it makes a number of assumptions, but effective).

Finally, Chapter 8 presents an overview of many common laser systems, including a summary of parameters (many of which are used in examples throughout the text). For the reader curious about the physical implementation of many lasers (e.g., "why does an argon-ion laser have a ceramic plasma tube?" or "why is the flashlamp so big on a ruby?"), this section attempts to answer a few basic questions on the design of these lasers.

I wish to thank Niagara College, Canada, for their cooperation in producing this text. With a dedicated photonics program, the labs at Niagara College have afforded me the ability to test numerous models on real lasers. Most of the photos in this text were shot at the college labs which feature an enormous range of lasers. I would also like to thank Ashley Gasque, acquiring editor, and Amber Donley, production coordinator, at Taylor & Francis, for their assistance.

Mark Csele
Welland, Ontario

The Author

Mark Csele is a full-time professor at Niagara College, Canada, in the heart of the beautiful Niagara Region. A physicist and a professional engineer, he has taught for over twenty years in programs ranging from a two-year technician level to a four-year undergraduate level. Currently, he teaches in the college's photonics programs, which feature an array of dedicated laboratories hosting a variety of laser systems. He has authored a previous book on fundamental laser concepts as well as several articles in magazines and trade encyclopedias.

1 Basic Laser Processes

In the course of developing models to predict laser performance, which is the focus of this text, we often need to start at the beginning. The purpose of this chapter, then, is to provide a basic overview of processes and parameters relevant to the laser. Some readers may already be familiar with many of these concepts but may not have seen them applied specifically to laser systems.

Key concepts outlined in this chapter include an atomic view of the laser processes of emission and absorption (including the important process of stimulated emission), a basic overview of rate equations involved in various laser levels (which will be useful when developing models later), methods of pumping a laser amplifier, and the nature of gain and loss in a laser amplifier. Only the basics are presented here and many of these concepts will be expanded upon in later chapters as required.

1.1 THE LASER AND LASER LIGHT

At first glance, a laser seems simple enough: an amplifier (often a glass tube filled with gas, a semiconductor diode, or a rod of glass-like material) pumped by some means, and surrounded by two mirrors—one of which is partially transmitting—from which an output beam emerges. As we shall see in this text, the arrangement is anything but haphazard.

Consider first the common helium-neon (HeNe) laser, which will be one of several lasers used extensively in examples in this text. Once the most popular visible laser, it has been replaced in many applications by smaller and cheaper semiconductor diode lasers. Still, the high-quality beam lends itself well to a host of laboratory applications (and, from the standpoint of an example laser, it represents a design with very "standard" elements. The structure of a typical HeNe laser is shown in the annotated photograph of Figure 1.1.

The HeNe laser is typical of many lasers: an amplifying medium surrounded by two cavity mirrors. The actual amplifying medium, in this case, is a mixture of helium and neon gases at low pressures and excited by a high-voltage discharge. In Figure 1.1, the actual discharge is seen to be confined to a narrow tube inside the larger laser tube and occurs between the anode (on which the output coupler, or OC, from which the output beam emerges is mounted) and the cylindrical cathode. (Note that not all tubes have the OC mounted on the anode; this varies by tube and may also be on the cathode end.)

Lasers are classed by the amplifying medium employed and may use gas, liquid (dye), solid-state (i.e. glass-like crystals), or semiconductor materials. Each has a

1

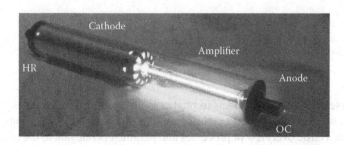

FIGURE 1.1   A basic HeNe gas laser.

special set of characteristics and a different method of pumping energy into the system. (Most solid-state lasers, for example, are optically pumped by an intense lamp source or another laser, whereas gas lasers are usually pumped by an electrical discharge through the gas medium.).

In a practical laser, a single photon of radiation passes through the amplifier many times, with the flux of photons becoming more powerful on each pass and eventually building to power where a usable beam results. The cavity mirrors surrounding the amplifier are required to produce lasers of a manageable length. If length were not a concern one could produce a HeNe laser tube over 80m in length and forgo mirrors altogether—but this is hardly practical and the use of mirrors allows a "folding" arrangement where photons are made to pass through an amplifier many times.

The required reflectivity of the cavity mirrors depends on the gain of the amplifier employed. Low-gain amplifiers require high-reflectivity mirrors—an example being the HeNe gas laser, which commonly has a high reflector (HR) of almost 100% reflectivity and an OC of about 99% reflectivity (with 1% of the intra-cavity power exiting through this optic to become the output beam). On the other hand, higher-gain lasers optimally use lower reflectivity optics and feature a higher transmission of the output coupler. Chapters 3 and 4 address the issue of optimal coupling.

The output of a laser is quite unique, as evident from simple observation. The most important property of laser light is coherence: every photon in the beam is in phase with each other. This is a consequence of the mechanism of the laser itself, stimulated emission, which ensures that all photons produced are essentially clones of an original photon—they must therefore all be of the exact same wavelength and relative phase.

Another important property of laser light is directionality and the extraordinarily low divergence of many laser beams. This property is primarily a consequence of the arrangement of optical elements in the laser. Most lasers are standing-wave lasers which have the general form of an amplifier surrounded by two cavity mirrors aligned very parallel to each other (as we shall see later, though, this is not the only possible arrangement). Photons emitted from the front of the laser will have made many passes through the amplifier (hundreds, or perhaps thousands) and so, in order to be amplified at all they must be on a specific trajectory perfectly aligned to the optical axis of the laser. Photons that are divergent, even slightly, will often strike the side of the amplifier tube and hence never be amplified to any extent.

1.2   ATOMIC PROCESSES OF THE LASER

In an incoherent light source, emission occurs as a result of electrons at an excited state relaxing to a lower energy level. The excitation source can be thermal, optical (absorption of a pump photon), or collision with a high-energy electron (as, for example, in a gas discharge). Once excited, atoms remain at an excited state for an average time that depends on the specific level involved (called the spontaneous lifetime), eventually falling to a lower level. The energy difference between the upper and lower levels of the transition is manifested as an emitted photon.

Incoherent light sources, then, utilize two atomic processes of interest: absorption and spontaneous emission. In a laser, a similar process to spontaneous emission occurs except the emission is stimulated to occur by a "seed" photon, which is "cloned" to produce two identical photons. Since the two resulting photons are identical (in wavelength and in phase), the resulting radiation is coherent. The process continues as photons traverse the amplifier producing a cascade of photons. This stimulated emission process is the key to laser action (the *SE* in the acronym *laser*).

Stimulated emission is actually the opposite process to absorption. As shown in Figure 1.2, absorption is a stimulated process: an atom at the lower level is stimulated, by absorbing a photon, to jump to the upper level. Similarly, stimulated emission requires an initial photon: an atom already at the upper energy level (through pumping) is stimulated to produce a photon by proximity to an existing photon—in the course of the process the original photon is not destroyed but rather passes through the medium with an identical clone produced in the process.

As we shall see in Section 1.4, the population of specific atomic levels is of interest to all processes. In a thermally excited source—for example, an incandescent lamp—these populations follow a Boltzmann distribution which predicts the population as:

$$N = N_0 e^{\frac{-E}{kT}} \qquad (1.1)$$

where N_0 is the total number of atoms in the system, T is the temperature in Kelvin, E is the energy of the level (in Joules) above ground state, and k is Boltzmann's constant.

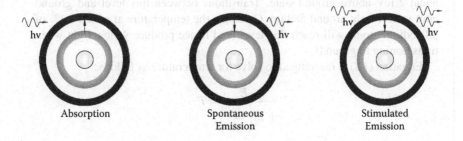

FIGURE 1.2   Atomic processes in a laser.

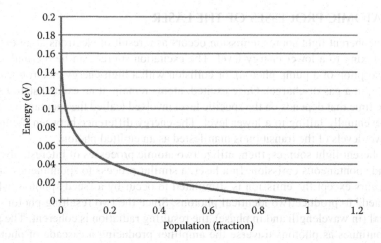

FIGURE 1.3   Boltzmann distribution of atomic populations.

Most conveniently, the equation can be rearranged to solve for the fraction of atoms reaching a certain level as:

$$\frac{N}{N_0} = e^{\frac{-E}{kT}}$$

(1.2)

This equation will be used later in the text when considering the thermal population of, for example, the lower level of transitions that are particularly close to ground level since this affects laser action.

Depending on the medium, energy levels may be discrete, well-defined levels or may be part of a "band" of energies (as with most metals). Assuming a continuous range of energy levels is available in a medium, the distribution of populations of atoms at each level resembles the curve of Figure 1.3.

EXAMPLE 1.1   EMISSION OF THERMAL LIGHT

Sodium has its first energy level (actually, a pair of levels closely spaced) at about 2.1eV above ground state. Transitions between this level and ground emit yellow light around 589nm. Calculate the temperature at which 0.1% of all sodium atoms will reach this level (and hence produce yellow light when transitioning to ground).

Equation (1.2) is rearranged to solve for temperature as follows:

$$\frac{-E}{k \; ln\left(\dfrac{N}{N_0}\right)} = T$$

Even in a system with discrete levels such as a gas discharge, levels usually fill from the "bottom up" and so lower levels will always have a larger population than upper energy levels. Such is the situation in most incoherent sources of light encountered in everyday life, where the upper level of a transition has a lower population than the lower level. Light production still occurs, of course, but in a system consisting of only a few energy levels absorption will dominate (since there are more atoms at the lower level available to absorb radiation). In this situation, amplification of radiation is not possible (this will be proven mathematically later in this chapter). For amplification to occur, emission must exceed absorption on a transition and so an inversion is necessary in which upper levels have a higher population than lower levels or, more specifically, the population of the ULL (upper lasing level, the level on which the lasing transition originates) must exceed that of the LLL (lower lasing level, the level on which the lasing transition terminates).

An inversion is hardly a "natural" state and must be coaxed into existence by the process of pumping which is, in a laser, the process of ensuring an inversion exists by delivering atomic populations to the ULL directly, bypassing the LLL (population of the LLL is detrimental to laser operation as it reduces available inversion and hence the gain of the amplifier).

As an example of how pumping is accomplished, consider the pumping mechanism of the helium-neon gas laser in which helium serves to transfer energy to a specific level in neon, where two levels serve as levels for the laser transition. The HeNe gas laser, with atomic levels outlined in Figure 1.4, consists of a mixture of helium and neon gases in the approximate ratio 10:1. The large proportion of helium gas will preferentially absorb energy from electrons traversing the discharge (instead of colliding with the less abundant neon atoms), pumping these helium atoms to an upper level situated 20.61eV above ground (arrows on the figure indicate upward and downward transitions in energy levels). Helium, a simple atom with only two electrons, has very few energy levels and by knowing the electric field applied to the tube (i.e. the voltage across the discharge and the length of the discharge itself), gas pressure can be adjusted to ensure that, on average, helium atoms are pumped preferentially

---

And the following values are substituted:

$$\frac{-2.1eV \times 1.602 \times 10^{-19} J/eV}{1.38 \times 10^{-23} k \; ln(0.001)} = T$$

The energy 2.1eV is first converted to Joules ($3.36*10^{-19}$J) as required, and the required temperature is found to be 3529K. High temperatures such as this are required for emission from any metal with "red hot" output from a metal beginning around 800K. At this temperature, only a tiny fraction of all atoms achieve an energy level suitable for emission of such radiation.

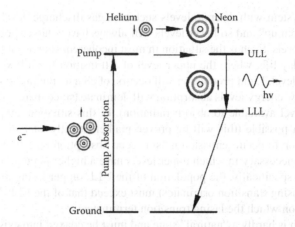

FIGURE 1.4 Pumping in a HeNe laser.

to the 20.61eV level as opposed to any other available level in helium (adjusting the pressure will set the mean-free path of electrons). Excited helium atoms then collide with neon atoms (which are considerably more massive than helium) resulting in inelastic collisions which transfer energy to the neon atom pumping them to the $5s^1$ level at 20.66eV above ground. (The normal electron configuration of neon is $1s^22s^22p^6$ and so there are a number of possible energy levels between this ground state and the excited state configuration of $1s^22s^22p^55s^1$ that we are discussing here.) It is a fortunate coincidence that helium (which has only a few energy levels) has an energy level at 20.61eV above ground state. This is very close to the 20.66eV energy level of neon, which serves as the upper level for visible transitions in this laser: when excited helium collides with neon in the ground state, a transfer of energy occurs in which the helium atom loses energy and neon is hence excited with the small difference between the 20.61eV from the helium atom and the 20.66eV level of the neon atom (i.e. the slight "upward" energy difference between the helium pump state and the neon ULL) filled by thermal energy present in the system (0.05eV is a very small energy difference). Pumping, then, is accomplished by careful design of the physical elements of the tube including gas pressure and ratio.

Similar pumping mechanisms exist in other gas lasers. In the carbon-dioxide gas laser, for example, electrons drive nitrogen gas molecules to an excited energy state from which it pumps populations of carbon dioxide molecules to a specific upper level by collision. In both the HeNe and CO_2 lasers, the pump level and the ULL reside in different atomic or molecular species.

The pumping method of various lasers varies. Semiconductor lasers, for example, are the most common type of laser and are small devices composed of a junction between p- and n-type semiconductor materials (i.e. a diode). Application of an electric field across the device causes charge carriers (electrons and holes) to be injected into the depletion region (a region of neutral charge formed where the semiconductor materials meet). Recombination of these charge carriers results in the

FIGURE 1.5   Semiconductor (diode) laser.

emission of photons. The actual pump source, then, is the injection of charge carriers accomplished by an electric current. Another method of pumping a semiconductor laser is optically, although this method is considerably less common.

Figure 1.5 shows a typical semiconductor diode laser. The actual device is only a few tenths of a millimeter in length and is mounted on a metal heatsink which also serves as an electrical contact. (For reference, consider that the "can" structure surrounding the window in the device through which the laser is visible is only 6mm in diameter.) On the top of the device, four thin gold wires visible in the photograph make electrical contacts between the actual semiconductor and a terminal that protrudes through the rear of the mount through which connection is made.

Solid-state lasers such as ruby (the first laser discovered), Nd:YAG, and others are optically pumped by either an arc or flashlamp, or by a diode laser in what is called a diode-pumped solid-state (DPSS) laser. Flashlamp pumping is required for some lasers (such as the ruby) and is an option for high-power pulsed operation in others (such as the Nd:YAG). Flashlamps emit very high peak powers well suited for some types of solid-state lasers, but while peak powers are large, overall efficiency is low given that such lamps convert at most a few percent of the total electrical input to light that is absorbed by the solid-state laser medium itself. Consider a solid-state laser pumped by a flashlamp. These lamps (similar to a photographic strobe lamp) are typically about 5% efficient at converting electrical energy into broadband optical radiation. Of that portion of radiation actually created by the lamp, only about 25% is actually emitted at wavelengths where it can be absorbed by the solid-state amplifier (ruby, for example, absorbs pump radiation only in the blue and green regions of the spectrum). Factor in coupling efficiencies (i.e. the portion that is actually transferred from the lamp to the solid-state amplifier), and the overall efficiency of the system is under 1%—but while this may seem low, flashlamps have enormous peak output

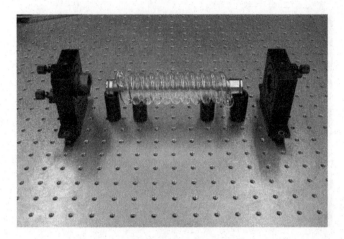

FIGURE 1.6   Components of a flashlamp-pumped solid-state laser.

powers (hundreds of kilowatts or more are common) unachievable by other pumping methods.

Figure 1.6 shows the major components of a solid-state laser (in this case, a ruby laser) assembled, for clarity, on a bench. Two cavity mirrors, in kinematic mounts allowing precise alignment, surround the 15cm long amplifier rod, which lies within a large helical flashlamp. In reality, the laser would be packaged (which would obscure the view of these components) with a reflector around the flashlamp as well as a water cooling jacket around both the lamp and the amplifier rod. (One might notice that the ends of the amplifier rod are embedded into metal mounts with O-rings, allowing the entire rod to be immersed in the same cooling enclosure as the lamp.)

Aside from the peak power delivered by a flashlamp for a pulsed laser, continuous wave (CW) solid-state lasers are most often pumped by diode lasers, which are very efficient given that they convert a large fraction of electrical energy into light, and furthermore, a diode can be chosen (or sometimes "tuned") so that the wavelength of emission corresponds well with an absorption peak of the solid-state laser medium.

Figure 1.7 shows the components of a compact DPSS unit with the completed assembly in the background. The main components consist of a pump diode (seen with two electrical leads protruding from the rear of that device), a tiny vanadate crystal (neodymium-doped ortho-vanadate or $Nd:YVO_4$) mounted on a copper disk, and an output coupler (OC) cavity mirror. The vanadate crystal, about 1mm thick, has one cavity mirror (called the high-reflector or HR) deposited directly onto the rear surface through which 808nm radiation from the pump diode passes to excite neodymium atoms in the medium. The OC is separate on this laser allowing optical alignment which, as we shall see, is crucial for laser operation. The output from this laser occurs at 1064nm and is of very high quality both in terms of beam optical quality and spectroscopically (for the spectral width of the output is very narrow). The laser shown in Figure 1.7 is actually somewhat unusual in the fact that output

FIGURE 1.7 DPSS laser components.

is in the IR—many small DPSS lasers of this type emit in the green at 532nm by frequency-doubling the 1064nm radiation using non-linear optical processes (which are covered in Chapter 7). A laser designed for green output incorporates a second crystal to perform this conversion.

While the DPSS laser pictured here is a small unit with an output power of under 10mW, DPSS lasers are scalable to enormous power levels of well over a kilowatt. While small DPSS lasers such as the one pictured are end-pumped along the optical axis of the laser via a single large pump diode, larger DPSS lasers are usually side-pumped from an array of many diodes (in much the same way as the flashlamp of Figure 1.6 pumps the solid-state material from the sides of the rod).

Lasers may be pumped in a variety of ways. Where the pump source is continuous, a continuous inversion may be possible (although not always so, as covered in Section 1.3), and where a pump source is pulsed, such as a flashlamp, inversion is a transient event and laser action can ensue only during the time when inversion is available (and further to that, as we shall see in Chapter 2, only when inversion is large enough to generate an optical gain sufficient to overcome optical losses in the laser system).

1.3 THREE- AND FOUR-LEVEL SYSTEMS

With only two atomic levels in a system, inversion cannot be achieved: photons will be absorbed as quickly as they are produced and, at best, a "break-even" situation will exist in which half of the atomic population will exist at the upper level and half at the lower level (and even that requires essentially infinite pump energy to accomplish). Practical lasers involve at least three levels such that pumping occurs (for this example, in an optically pumped laser) at a shorter wavelength and emission at a longer wavelength preventing the lasing transition from absorbing the majority of photons produced in the laser.

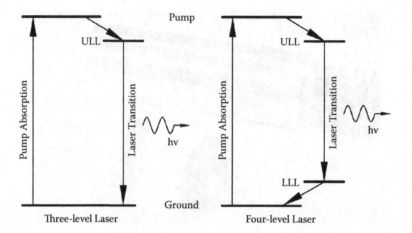

FIGURE 1.8 Three- and four-level lasers.

Lasers are characterized as three- or four-level system based on the presence, or absence, of a discrete LLL. The basic energy levels for idealized examples are illustrated in Figure 1.8. A three-level system is characterized by the lack of a discrete LLL, with the ground state serving this purpose. Several issues arise because of this configuration, with the issue here centered on the lifetime of ground state: infinity! The lifetime of the LLL always exceeds that of the ULL in a three-level laser and so inversion cannot be maintained continuously. Laser action is still possible, of course (the world's first laser, the ruby, is a three-level laser), but these lasers are pulsed. Assuming a long ULL lifetime, an inversion may be maintained for a period of time over which laser oscillation is possible. (In the case of ruby, the ULL lifetime is 3ms which allows ample time to fire a flashlamp and build a large inversion—laser action must then ensue well before the lifetime expires since the definition of "level lifetime" is the point at which 50% of the population decays and so inversion will be lost by this point.)

As with all lasers, an inversion is required, and to generate this inversion in a three-level laser an enormous amount of pump energy is required since over 50% of the total quantity of atoms in the amplifier must be pumped from the ground level (which itself has an enormous population of atoms ... when not pumped all atoms will reside at this energy level). The minimum pump energy to achieve inversion for a three-level laser is hence:

$$E_{PUMP} = \frac{1}{2}(\# \ atoms)h\upsilon \tag{1.3}$$

This estimate will always be low and more energy is required since this energy does not really generate an inversion at all, just a break-even point where one-half of the atoms are at the ground state and one-half in the ULL. More energy will be required to generate an actual inversion, which is needed to produce optical gain (as we shall see in Chapter 2).

EXAMPLE 1.2   ACHIEVING INVERSION IN A THREE-LEVEL LASER

Consider a ruby laser (Cr^{3+} doped Al_2O_3) that is pumped by green light at approximately 500nm. The rod is 75mm long by 6.35mm in diameter and is doped with Cr^{3+} to 0.05% by weight.

The density of ruby is 3.98g/cc and the rod has a volume of:

$$V = l \times \pi r^2 = 2.375 cm^3$$

So the weight of the rod is 9.45g. The weight of chromium is hence 0.05% of this weight or 4.73mg. Knowing that chromium has an atomic mass of 52g, the rod is doped with:

$$N = 6.023 \times 10^{23} \, atoms/mol \times \frac{4.73 \times 10^{-3}}{52}$$

or 5.47*10^{19} ions (which represents 9.09*10^{-5} mol of chromium). Using Equation (1.3), the pump energy required to bring 50% of the atoms to the ULL is calculated to be:

$$E_{PUMP} = \frac{1}{2}(5.47 \times 10^{19}) \frac{hc}{500 \times 10^{-9} m}$$

or 10.88J of actual absorbed energy. A flashlamp is required for pumping a ruby laser, with most flashlamps sporting an efficiency of about 1% at converting electrical energy into actual usable wavelengths that are absorbed by the ruby material. The electrical input required, then, is 1088J.

This is an enormous amount of energy (considering the average "high power" photographic strobe operates at about 50J) and will require a very large pump lamp and energy storage system, not to mention a cooling system to remove excess heat—this helps illustrate the basic problem with three-level lasers—and while this energy is large it is insufficient to generate any real gain, though, and so additional pump energy will be required to bring the laser to oscillation.

In the case of a four-level laser, the addition of a discrete LLL to the system makes generation of an inversion much easier, since one no longer has to raise a full 50% of all atoms to the ULL. In fact, in a typical four-level laser only a small fraction of the total number of atoms available in the amplifier are ever involved in an inversion.

Under the right conditions, an inversion can be continuously maintained and so CW lasing action is possible, at least in a four-level system.

A four-level laser may have only three actual levels, as it is possible in some lasers to skip the pump level and pump energy directly into the ULL. These lasers are still classed as four-level lasers due to the presence of a discrete LLL. Similarly, some quantum systems have five or more levels, often with intermediate levels between the LLL and ground (the HeNe laser has such a level, a metastable energy level around 16.6 eV, as a necessary step in the depopulation of the LLL to ground) and again, these are classed as four-level lasers.

To further complicate matters, where the LLL of a laser system is particularly close to ground state it may become thermally populated (as predicted by the Boltzmann equation) in which case a quasi-three-level laser results. This specific case will be considered in Chapter 5.

1.4 RATE EQUATIONS

It is instructive to consider the basic rate equations governing laser operation, as several key conclusions result. We begin by considering the two levels actually involved in the production of laser radiation: the upper level, now called the upper lasing level or ULL, and the lower level, now called the lower lasing level or LLL. Between these two levels three processes of interest occur as outlined in the previous section: absorption, spontaneous emission, and stimulated emission. Absorption is a stimulated process requiring the presence of radiation to occur. In general, the rate of absorption can be expressed as:

$$r_{ABSORPTION} = N_{LLL}\rho B_{12} \tag{1.4}$$

where N_{LLL} is the population of the lower level from which absorption occurs, ρ the photon flux in the laser, and B_{12} the Einstein coefficient (a proportionality constant) for a transition from level 1 to level 2 (hence the subscript "12"). We may also express this in terms of the probability of absorbing a pump photon:

$$W_P = \rho B_{12} \tag{1.5}$$

And so the rate of absorption becomes:

$$r_{ABSORPTION} = N_{LLL}W_P \tag{1.6}$$

This makes sense: the rate of absorption is proportional to the number of atoms at the lower level available to absorb photons (N_{LLL}) as well as the number of photons available to absorb (embodied in probability W_P).

Spontaneous emission is a simple, non-stimulated process with a rate defined by:

$$r_{SPONTANEOUS} = \frac{N_{ULL}}{\tau} \tag{1.7}$$

where τ is the lifetime of the upper level (and so $1/\tau$ is the rate at which the population of this level decays—this is a common physical parameter that can be researched). The third process, stimulated emission, is similar to absorption:

$$r_{STIMULATED} = N_{ULL}\rho B_{21} \tag{1.8}$$

This rate is dependent on the population of the upper level, the photon flux, and the Einstein coefficient of stimulated emission (with the subscripts reversed to signify that it applies from levels 2 to 1, i.e. the upper level to the lower level or emission).

In a laser, the rate of stimulated emission must exceed the rate of absorption (for the levels of the lasing transition) so that the amount of radiation produced by the laser exceeds the amount absorbed. One may form a simple ratio between these two rates (Equations 1.8 and 1.4), and assuming that the Einstein constants for absorption and emission are the same, it may be concluded that to keep these rates in the desired ratio (i.e. stimulated emission greater than absorption) an inversion (with the population of the ULL greater than that of the LLL) is required.

The above rate equations express the rates of the individual processes and we will now use these to express the rate of individual levels—for example, the rate at which the population of the ULL or the LLL changes. This treatment is useful when considering, for example, the amount of pump power required to generate an inversion (as will be done in future chapters).

Consider, first, the pump level:

$$\frac{dN_{PUMP}}{dt} = IN - OUT = r_{ABSORPTION} - r_{DECAY} \tag{1.9}$$

where $r_{ABSORPTION}$ is the rate at which pump photons are absorbed to populate the pump level and r_{DECAY} is the rate at which the population of that same level decays to the ULL. Substituting the rates previously developed this rate equation becomes:

$$\frac{dN_{PUMP}}{dt} = N_{LLL}W_{PUMP} - \frac{N_{PUMP}}{\tau_{PUMP}} \tag{1.10}$$

The second term is simply spontaneous decay from the pump level to the ULL and the time constant is specific to that transition. Aside from decay from the pump level to the ULL, other undesirable transitions are possible, for example between the pump level and the lower lasing level. For an efficient laser those time constants must be far larger than that for the pump to the ULL (so that it is far less probable and hence the majority of atoms decay to the ULL to populate that level). For that reason, we will assume in this "perfect" laser that pump level always decays to the ULL (and so the lifetime of the level is the same as the lifetime of the level with respect to this specific transition).

We have also taken the liberty here of simplifying Equation (1.10) by the use of N_{LLL} as the number of atoms available to pump. To be more accurate, one should express the number of available atoms as $(N_{LLL} - N_{PUMP})$; however, in a "good" laser the lifetime of the pump level is very, very short and so an appreciable population of atoms never builds in this level: atoms entering the pump level are assumed to decay almost immediately to the ULL.

The rate equation for the ULL can be expressed as:

$$\frac{dN_{ULL}}{dt} = \frac{N_{PUMP}}{\tau_{PUMP}} - \frac{N_{ULL}}{\tau_{ULL}} - N_{ULL}\rho B_{21} \qquad (1.11)$$

where the first term is the rate at which population enters the ULL (which is, of course, the rate of decay from the pump level), the second term is the rate of spontaneous emission between the ULL and the LLL, and the final term the rate of stimulated emission. In the absence of intra-cavity radiation (i.e. before threshold is reached and laser output appears) the intra-cavity photon flux can be considered negligible and so the equation for the ULL can be simplified to:

$$\frac{dN_{ULL}}{dt} = \frac{N_{PUMP}}{\tau_{PUMP}} - \frac{N_{ULL}}{\tau_{ULL}} \qquad (1.12)$$

At this point, we proceed to develop an expression for inversion. In a steady-state condition, the rate of population growth in the pump level and the ULL are both zero, allowing a few fundamental substitutions to be developed. In the first case, the rate of the pump level (1.10) is zero:

$$\frac{dN_{PUMP}}{dt} = N_{LLL}W_{PUMP} - \frac{N_{PUMP}}{\tau_{PUMP}} = 0 \qquad (1.13)$$

and so we can express the population of the pump level as:

$$N_{PUMP} = N_{LLL}W_{PUMP}\tau_{PUMP} \qquad (1.14)$$

In a similar manner, the rate of the ULL (1.12) is set to zero:

$$\frac{dN_{ULL}}{dt} = \frac{N_{PUMP}}{\tau_{PUMP}} - \frac{N_{ULL}}{\tau_{ULL}} = 0 \qquad (1.15)$$

allowing an expression of the population of the ULL as:

$$N_{ULL} = N_{PUMP}\frac{\tau_{ULL}}{\tau_{PUMP}} \qquad (1.16)$$

We may now use a general expression for inversion,

$$\Delta N = N_{ULL} - N_{LLL} \qquad (1.17)$$

and substitute for N_{ULL} using Equation (1.16) and then Equation (1.14):

$$\Delta N = N_{PUMP}\frac{\tau_{ULL}}{\tau_{PUMP}} - N_{LLL} \qquad (1.18)$$

$$\Delta N = N_{LLL}W_{PUMP}\tau_{ULL} - N_{LLL} \qquad (1.19)$$

$$\Delta N = N_{LLL}(W_{PUMP}\tau_{ULL} - 1) \qquad (1.20)$$

This expression is valid for a three-level laser and three major ramifications result. First, to have a positive inversion (i.e. to have an inversion at all), a significant rate of pumping is required such that $W_{PUMP}\tau_{ULL} > 1$. Second, the longer the lifetime of the ULL, the lower the rate of pumping required to make this inversion (since there is more time to build the inversion), and so in a three-level laser a long ULL lifetime is very desirable and perhaps even required to make inversion possible. And third, for a three-level laser LLL is ground and so N_{LLL} is a very large number meaning that a massive amount of energy is required to get to this break-even point. The reader is cautioned, though, that reaching a positive inversion will not allow the laser to oscillate—the inversion must now build (through additional pumping) to produce enough gain to overcome losses in the laser (of which many are unavoidable).

Analysis of a four-level laser proceeds in much the same way as the three-level laser. The basic rate equations for the pump (1.10) and ULL (1.12) hold true as well as the expressions for the population of the pump (1.14) and ULL (1.16) already developed. In a four-level laser, the LLL is a discrete level and so has a rate equation of:

$$\frac{dN_{LLL}}{dt} = \frac{N_{ULL}}{\tau_{ULL}} - \frac{N_{LLL}}{\tau_{LLL}} \tag{1.21}$$

This allows an expression for the population of the LLL:

$$N_{LLL} = N_{ULL}\frac{\tau_{LLL}}{\tau_{ULL}} \tag{1.22}$$

Inversion in a four-level laser can hence be expressed as per (1.17) and substitution made for the ULL population by (1.16) and the LLL by (1.22):

$$\Delta N = N_{PUMP}\frac{\tau_{ULL}}{\tau_{PUMP}} - N_{ULL}\frac{\tau_{LLL}}{\tau_{ULL}} \tag{1.23}$$

And further expansion of N_{ULL} by (1.16):

$$\Delta N = N_{LLL}W_{PUMP}\tau_{ULL} - N_{LLL}W_{PUMP}\tau_{LLL} \tag{1.24}$$

We now realize that for an ideal laser the lifetime of the ULL is much longer than the lifetime of the LLL. (In the next section we calculate the lifetimes for real lasers and indeed see that the lifetime of the ULL is often ten times longer than that of the LLL for many practical lasers.) The second term in Equation (1.24) above is hence much smaller than the first, becoming insignificant, and so the expression for inversion simplifies to:

$$\Delta N = N_{LLL}W_{PUMP}\tau_{ULL} \tag{1.25}$$

Even if the LLL lifetime is considered to be finite and significant, the lifetime of the ULL will always be longer in a "good" four-level laser and so inversion in a four-level laser builds immediately with the onset of pumping. Unlike the three-level laser, there is no threshold of pumping just to achieve inversion. Of course, an inversion must be built up large enough so that the optical gain generated exceeds loss in the system (but unlike the three-level laser, we did not have to achieve a break-even population at the ULL first!). Four-level lasers are hence far more efficient than three-level lasers and so the vast majority of practical lasers are four-level, although we are not "done" with the three-level laser yet as some four-level lasers behave in a quasi-three-level manner due to a LLL close to ground state which becomes thermally populated at operating temperatures.

In this section, many simplifications have been made including the assumption that the ULL lifetime was far longer than the LLL lifetime. Had this not been true, oscillation might still be possible; however, a pulsed laser would result in which there would also be specific requirements for the pumping pulse (as there are with a three-level laser, namely that pumping must occur in a time frame of well under the ULL lifetime). In the next section these lifetimes will be calculated.

1.5   LEVEL LIFETIME

The ability of a particular system to operate in CW (continuous wave) mode (i.e. to produce a continuous laser output) is determined, in large part, by the lifetime of the upper- and lower-lasing levels. In order to maintain a continuous inversion the lifetime of the ULL must exceed that of the LLL—if this criteria is not met, inversion (and hence lasing) may still be achieved; however, inversion will be lost by the time the lifetime of the ULL is reached and the laser will be self-terminating (and hence strictly a pulsed laser).

As an example, consider the molecular nitrogen laser, which is self-terminating. In this particular laser the lifetime of the ULL is approximately 20ns while the LLL is 40μs in length. Assuming pumping can be supplied in a huge pulse in a time period of well under the ULL lifetime (which is quite possible, although this requires careful design), inversion can be achieved and lasing will begin; however, by the time τ is reached (20ns), 50% of all molecules in the ULL will have decayed to the LLL and inversion will be lost. Since this is a four-level laser, molecules in this LLL state are not available for lasing and must decay to ground first before being available to be pumped back to the ULL again. The result is a pulsed laser with pulse widths under 20ns (usually around 10ns), although repetitive pulses can be generated at a high rate.

Most common four-level lasers, including the HeNe, YAG, and most semiconductor diode lasers, feature a more optimal situation where the ULL lifetime exceeds that of the LLL and so inversion can be maintained continuously. Although the lifetime of a particular ULL, like most physical constants, can be found in published research or in tabulated form it is instructive to examine how this figure originates and so the process will be demonstrated here for the HeNe laser.

EXAMPLE 1.3   LIFETIME OF HENE ENERGY LEVELS

In order to consider an atomic transition for CW laser action, one must ascertain the lifetime of the ULL and the LLL. These computations lend themselves to other considerations as well; for example, in Chapter 6 the concept of Q-switching is considered, for which level lifetime plays an immense role in determining suitability of application of that technique. For this example, the lifetimes of the levels involved in the common HeNe laser are considered.

One begins with a physics reference such as the *NIST Atomic Spectral Database*,[1] which lists transition probabilities as well as the upper energy levels for each transition. Consider the red 632.8nm output of the HeNe laser—NIST data for this transition (Figure 1.9) shows this transition as having an upper energy level (a $5s^1$ level) of $166656.5 cm^{-1}$, or 20.66eV, which is circled in the figure.

Quickly scanning through other neutral neon transitions (with most verified using a secondary reference listing all known laser lines[2,3]) in the same table (as outlined in Figure 1.10) reveals many transitions originating from the exact same upper level, for example 611.8nm and 543.4nm. Careful examination of these tables reveals a total of ten transitions originating from this same level (many are able to lase as well, such as 611.8nm which is available commercially as an orange HeNe or 543.3nm as a green HeNe although none of these wavelengths is as common as the ubiquitous 632.8nm red laser).

Summarizing the data (since there are ten transitions originating from the same transition spanning many pages of data), the transition probabilities for each transition originating from the same ULL as the 632.8nm transition are summarized in Table 1.1.

Most transitions in Table 1.1 (all of the visible ones) represent those from the $5s^1$ to the $3p^1$ levels. The exception is the 3391nm IR transition, which terminates on the $4p^1$ level. There are other minor transitions, for example 2416nm (between this same ULL and a different $4p^1$ level), but these probabilities are

Observed Wavelength Vac (nm)	Ritz Wavelength Vac (nm)	Rel. Int. (?)	A_{kj} (s^{-1})	Acc.	E_i (cm^{-1})	E_k (cm^{-1})	Lower Level Conf., Term, J	Upper Level Conf., Term, J
632.81646	632.8164+	3000	3.39e+06	B	150 858.5079 – 166 656.5114		$2s^22p^5(^2P°_{1/2})3p\ ^2[3/2]\ 2$	$2s^22p^5(^2P°_{1/2})5s\ ^2[1/2]°\ 1$

FIGURE 1.9   NIST data for the 632.8nm HeNe laser transition.

Observed Wavelength Vac (nm)	Ritz Wavelength Vac (nm)	Rel. Int. (?)	A_{kj} (s^{-1})	Acc.	E_i (cm^{-1})	E_k (cm^{-1})	Lower Level Conf., Term, J	Upper Level Conf., Term, J
632.81646	632.8164+	3000	3.39e+06	B	150 858.5079 – 166 656.5114		$2s^22p^5(^2P°_{1/2})3p\ ^2[3/2]\ 2$	$2s^22p^5(^2P°_{1/2})5s\ ^2[1/2]°\ 1$
611.80187	611.8014+	150	6.09e+05	B	150 315.8612 – 166 656.5114		$2s^22p^5(^2P°_{3/2})3p\ ^2[3/2]\ 2$	$2s^22p^5(^2P°_{1/2})5s\ ^2[1/2]°\ 1$
604.61348	604.6133+	500	2.26e+05	B	150 121.5922 – 166 656.5114		$2s^22p^5(^2P°_{3/2})3p\ ^2[3/2]\ 1$	$2s^22p^5(^2P°_{1/2})5s\ ^2[1/2]°\ 1$
543.36513	543.36499+	2500	2.83e+05	B	148 257.7898 – 166 656.5114		$2s^22p^5(^2P°_{3/2})3p\ ^2[1/2]\ 1$	$2s^22p^5(^2P°_{1/2})5s\ ^2[1/2]°\ 1$

FIGURE 1.10   NIST data for a few transitions originating from the same ULL.

TABLE 1.1

Transition Probabilities of Neon Transitions Originating from the Exact Same Level as the 632.8nm Laser Transition

Wavelength (nm)	Transitional Probability ($10^8 s^{-1}$)
543.5	0.00283
594.1	0.00200
604.6	0.00226
612.0	0.00609
629.4	0.00639
632.8	0.0339
635.2	0.00345
640.3	0.0139
730.7	0.00255
3391	0.00414

quite insignificant and so are not included here. Finally, there is an eleventh transition between the ULL and the ground state; however, it is not significant here—it will be revisited in Chapter 3 as it has special characteristics.

To determine the lifetime of this specific $5s^1$ upper level, then, all transitional probabilities in Table 1.1 sum to A = $7.75*10^6$ s^{-1} and so the inverse (which is the effective lifetime of the ULL) is found to be 129ns.

One must now consider the lower level of the 632.8nm transition (a $3p^1$ level) and using the same transition tables, the lifetime of the LLL can also be calculated. Knowing the transition at 632.82nm has an ULL of 166657cm^{-1}, the wavelength 632.82nm is converted to units of cm^{-1} and the LLL is then found to be 150859 cm^{-1} (all energies are expressed in cm^{-1} since this is the given unit for energies in these tables).

Examining the tables again, three transitions are found to originate at the same level (150859 cm^{-1}), all of which serve to depopulate this level. These transitions are outlined in Table 1.2.

TABLE 1.2

Transition Probabilities of Neon Transitions Originating from the LLL of the 632.8nm Laser Transition

Wavelength (nm)	Transitional Probability ($10^8 s^{-1}$)
594.5	0.113
609.6	0.181
667.8	0.233

All three terminate on a 3s¹ level, which is logical given that the only level below this is the ground state which is a 2p level and basic quantum mechanics prohibits a transition between two "p" levels. The total A for those three transitions is found to be $0.527*10^8 s^{-1}$ and so the lifetime of the LLL is found to be 19ns. Hence the LLL is found to have a lifetime almost seven times shorter than the ULL and so it is concluded that CW laser action is indeed possible on this line. Similar calculations on other lines originating from the 5s¹ level of neon show that essentially all visible lines in the HeNe system originating from that same ULL can operate in CW mode.

To recap, key transitions and energy levels involved in this transition are outlined in Figure 1.11.

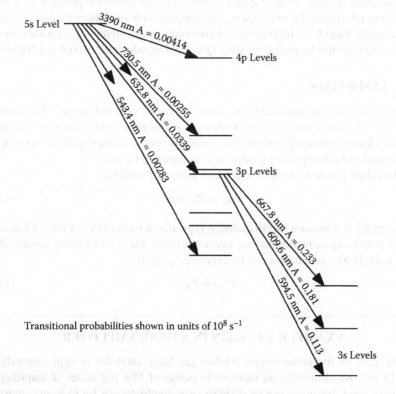

FIGURE 1.11 Neon transitions significant to the HeNe laser.

As well as determining the suitability of a system for CW lasing action, level life-time also has a role in determining the way in which a pulsed laser is pumped (regard-less of whether the laser is three- or four-level). In a pulsed laser, the flashlamp must logically produce all output within a time frame of less than the upper-lasing level (ULL) lifetime. If, for example, a particular laser medium has an ULL lifetime of 100μs, the pump source must logically pump and produce an inversion well before then since "lifetime" is defined as the point at which 50% of an atomic population is lost due to exponential decay. This places certain demands on the pump mecha-nism which, practically speaking, prohibits or limits the use of certain laser materials in certain configurations. As an example, consider the solid-state titanium-sapphire amplifier which has a short ULL lifetime of 3.8μs— when pumped by a flashlamp, this places the demand that a very fast lamp system that discharges in under a few microseconds be used. (The majority of these type of lasers are pumped by a laser other than a flashlamp for this reason, and pumping with a CW laser renders this laser CW as well). Short ULL lifetimes also have ramifications on the way in which energy in the amplifier may be used (including Q-switching, which is covered in Chapter 6).

1.6   LASER GAIN

The purpose of a laser amplifier is, of course, to provide gain and increase the power of a flux of photons traveling through the element. As photons travel through the amplifier they are cloned, producing more photons, which in turn are cloned again … the process is continual and power grows exponentially, as shown in Figure 1.12.

The actual power gain, then, as a rough estimate, would be:

$$P_{OUT} = P_{IN} \times gx \tag{1.26}$$

where gain g is a per-unit length quantity, typically in units of m^{-1} or cm^{-1}. Of course power grows exponentially and so, mathematically, the power exiting an amplifier after a single pass of length x can be expressed properly as:

$$P_{OUT} = P_{IN}e^{gx} \tag{1.27}$$

EXAMPLE 1.4   GAIN IN A HENE AMPLIFIER

The gain of the helium-neon (HeNe) gas laser amplifier is approximately 0.15 m^{-1}, which implies an increase in power of 15% per meter of amplifier length—not that many practical HeNe laser amplifiers are 1m in length; most lasers of this type are about 30cm. Using Equation (1.27) for a more accurate figure (since gain grows exponentially),

$$P_{OUT} = P_{IN}e^{gx} = P_{IN}e^{0.15m^{-1} \times 0.3m} = 1.046$$

Numerically, power grows 4.6% on every single pass through the 30cm HeNe laser amplifier. This may seem like a large amount, but many losses in the laser (as covered in Chapter 2) quickly consume this gain.

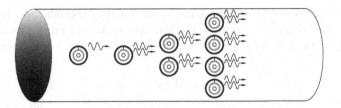

FIGURE 1.12 Exponential power gain.

One might think that by simply doubling the input power to an amplifier, gain will double, or perhaps that making an extraordinarily long amplifier would be a viable option; however, in the case of many laser amplifiers there are engineering or physical limits that prohibit this. In the HeNe laser, for example, there is a physical limit on how much power may be pumped into an amplifier before inversion, and hence gain actually decreases. And in the case of an extraordinarily long amplifier, practical limits exist on the length of a laser tube both due to alignment requirements (since a long laser presents a mechanical problem in terms of keeping the assembly stable) and because long gas discharge tubes require an excessively high voltage to operate. For these reasons the largest commercially built HeNe lasers are about 1m in length. Most other types of lasers feature some sort of limit on the gain available from the amplifier.

1.7 LOSSES IN A LASER

The laser system features a gain element (the amplifier) but inevitably features losses as well. Among other losses, the medium of a laser amplifier, whether gas, liquid, or solid, will have a finite amount of loss due to scattering and absorption. In addition to the amplifier, the cavity optics are a large source of loss. In many lasers, the HR is listed as "100% reflecting," but no mirror is absolutely perfect and so a small loss will occur there, and by definition, the output coupler represents a significant loss since it must allow a portion (usually between 1% and 10% for many common lasers) of the intra-cavity beam to pass through to become the output beam.

The significance of these losses is that the amplifier must produce enough gain, at a bare minimum, to overcome these since it is foremost an oscillator and requires positive feedback (so gain, practically, must exceed loss). Some losses are, unfortunately, intrinsic to the system and cannot be controlled or reduced: for example, scattering in a solid-state laser amplifier which, while it may be a feature of the purity or method of growth of the crystal, is generally beyond the control of the designer. Other parameters such as the reflectivity of optical elements can be controlled; thus if the gain of the amplifier is known, the optics can be designed appropriately.

Losses in the laser amplifier itself are characterized by the Greek symbol γ, which, like gain, is a per-unit length measure of loss along the entire length of the amplifier (usually measured in units of m^{-1} or cm^{-1}).

Transmission of a "lossy" optical element is a quantity described by:

$$T = e^{-\gamma x} \tag{1.28}$$

where γ is the per-unit length loss and x is the length of the element—in the case of a laser, x is usually the length of the amplifier itself over which loss (and gain) occur and not the distance between cavity mirrors. The actual loss incurred is then:

$$L = 1 - e^{-\gamma x} \qquad (1.29)$$

Both loss and gain are examined in Chapter 2, in which equations will be formulated to describe the threshold of a laser system as a whole.

1.8 CAVITY OPTICS

Various cavity structures are possible for a laser system, including stable and unstable configurations. A stable cavity is one that traps radiation such that it is reflected continually through the amplifier until it exits through the OC, while an unstable cavity allows radiation to escape, often around an optic. The amplification factor of most lasers is quite modest and so most lasers require a stable cavity configuration.

Figure 1.13 shows several common cavity configurations. Plane-parallel, while the simplest cavity and one that utilizes a large volume of the amplifier, is only marginally stable and so alignment (and maintaining alignment despite the fact that minor changes in temperature will result in the expansion and contraction of the frame holding these mirrors) is very difficult to achieve and maintain. The confocal arrangement is the easiest to align, since the foci for each cavity mirror lies directly on the surface of the other mirror, but usage of the amplifier volume is not optimal. Hemispherical cavities allow the production of a small beam waist (shown by arrows

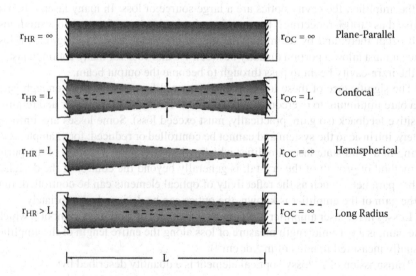

FIGURE 1.13 Laser cavity configurations.

on the figure), but usage of the amplifier volume is poor. A popular variation of the hemispherical arrangement is the long-radius cavity in which the concave mirror has a radius of curvature longer than the cavity. This arrangement, popular with many common commercial lasers, is a tradeoff between efficient usage of the amplifier volume and the ability to align the cavity properly.

Stability is determined by the radii of each mirror as follows:

$$0 \leq \left(1 - \frac{L}{RC_1}\right)\left(1 - \frac{L}{RC_2}\right) \leq 1 \qquad (1.30)$$

where L is the length of the cavity (i.e. the distance between the two mirrors), and RC_1 and RC_2 are the radii of curvature of each mirror.

EXAMPLE 1.5   STABILITY OF A HENE CAVITY

A HeNe laser has a plano-concave cavity with one flat mirror and one concave mirror with a radius-of-curvature of 0.5m. Calculate the spacing of the cavity mirrors for a stable cavity.

A flat mirror has a radius-of-curvature of infinity and so we solve for L in Equation (1.30), substituting for both mirror radii at the boundary conditions of zero and one:

$$0 \leq \left(1 - \frac{L}{\infty}\right)\left(1 - \frac{L}{0.5m}\right) \leq 1$$

Solving for cavity length L at the first boundary condition,

$$0 = 1 - \frac{L}{0.5m}$$

So $L = 0.5$m in this case. Similarly, solving for L at the second boundary condition,

$$1 = 1 - \frac{L}{0.5m}$$

with $L = 0$ in this case. The separation of the cavity mirrors can hence be any-where between zero and 0.5m to yield a stable cavity configuration.

Assume now that two concave mirrors are used (e.g. two concave HRs), both of 0.5m. Equation (1.30) can once again be solved (this time with RC_1 and RC_2 both equal to 0.5m) for cavity length which will be found to have limits at zero and 1m.

Although we have used boundary conditions (zero and one) to compute limits here, designing a cavity at either of these limits yields a marginally stable cavity and is not recommended, as any change in the cavity dimensions or alignment (for example, due to thermal expansion) will cause a change in the output (the laser may even cease oscillation). A plane-parallel configuration using two flat mirrors is such a cavity. Often, one mirror is made to be of a long radius to improve stability.

Aside from those shown, many other cavity arrangements are possible including unstable cavities in which radiation is allowed to escape the cavity (for example, through a hole in the OC or around the perimeter of the OC). An unstable cavity is often designed such that an intra-cavity photon generally makes a known number of passes through the amplifier before escaping. Such cavities are used with high-gain lasers.

The divergence of a laser beam is largely determined by the optical cavity of a laser and by the dimensions, since diffraction can occur where an aperture approaches the dimensions of a wavelength (as is the case with a diode laser). Semiconductor diode lasers have a high divergence which is a feature of the structure of the amplifier. Two basic types of diode structures exist for laser diodes: gain-guided and index-guided. Gain-guided lasers restrict the amplifying medium by the shape of the electrode on the top of the device: where current tapers off away from the electrode, gain decreases.

Index-guided structures feature the semiconductor junction within a groove in the device (which is the actual amplifying region—see Chapter 8 for more details). The change in refractive index between the material of the junction and that of the basic substrate in which the groove is manufactured results in a reflection similar to that in a fiber optic. The amplifying medium, then, acts as a waveguide. Regardless of the structure employed, dimensions of the active area result in an elliptical beam that diverges more in one direction than the other, as shown in Figure 1.14. To form a round "beam" from such a laser requires collimating optics that reshape the beam, such as use of two anamorphic prisms.

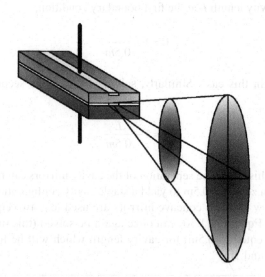

FIGURE 1.14   Beam shape of semiconductor laser.

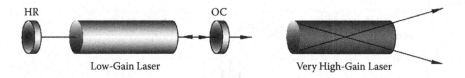

Low-Gain Laser   Very High-Gain Laser

FIGURE 1.15 Divergence in a superradiant laser.

Still other amplifying media have an enormous gain such that a single photon is amplified into a powerful stream of photons in only a single pass (with some amplifiers being smaller than a centimeter in length, indicating just how enormous the gain really is). These are called "superradiant" amplifiers. One drawback to superradiant lasers is, generally speaking, a highly divergent beam and a wide spectral width results as opposed to a laser utilizing a lower-gain medium. Since most superradiant lasers require little or no feedback (some require no mirrors whatsoever to oscillate), many photons originating at the rear of the laser amplifier on a variety of trajectories will be amplified, whereas with a low-gain medium only photons in a trajectory directly along the optical axis of the amplifier will be able to pass through the amplifier many times and hence form the output beam (which will then be highly collimated). The origins of this divergence are outlined in Figure 1.15.

1.9 OPTICAL CHARACTERISTICS (LONGITUDINAL AND TRANSVERSE MODES)

Most lasers, but not all, feature a cavity consisting of two mirrors surrounding an amplifier. This cavity is, in effect, an interferometer and so supports standing waves inside the cavity with a node of zero amplitude required at each mirror, as diagrammed in Figure 1.16. Where two plane mirrors are used in the cavity and are parallel to each other, the configuration is that of a Fabry-Perot interferometer.

In a cavity such as this, an integral number of half-wavelengths must fit within the cavity giving rise to discrete longitudinal modes separated by a constant frequency called the free spectral range (FSR) of the cavity.

$$FSR(Hz) = \frac{c}{2nl} \tag{1.31}$$

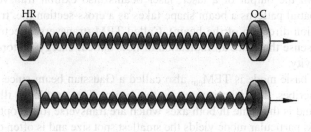

FIGURE 1.16 Longitudinal modes in a laser cavity.

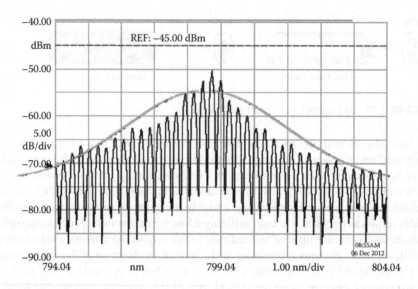

FIGURE 1.17   Expected and actual laser output.

where n is the index of refraction and l is the cavity length. Couple this periodic behavior with the fact that the gain profile of practical laser amplifiers takes the form of a Gaussian profile (i.e. gain is "spread out" over a range of wavelengths according to a Gaussian profile with a peak gain in the center of the profile), and the actual output of a laser will be seen to consist of a series of longitudinal modes (this is sometimes called *multimode* but this is somewhat ambiguous since, as we shall see later in this section, one must first define a "mode" as being transverse or longitudinal). The output from an actual semiconductor laser is illustrated in Figure 1.17 in which many modes are visible. The expected gain profile of the same laser, a Gaussian profile, is drawn as an overlay on the same figure. Note that at all points where the laser oscillates at all, gain exceeds losses in the laser—the actual gain profile is wider than shown in the figure. The appearance of FSR peaks in the output spectrum of a laser will be exploited later, in Chapter 2, to measure the actual length of a laser cavity optically.

In addition to longitudinal modes, which are in reality differences in discrete frequencies in the output of a laser, laser beams also exhibit transverse modes which are spatial patterns a beam shape takes as a cross-section (i.e. transverse to the propagation direction of the beam). Called TEM (transverse electromagnetic) modes, in a sense these are an indication of how optical energy is stored within a particular cavity.

The most basic mode is TEM$_{00}$, also called a Gaussian beam since a Gaussian function describes the intensity across the beam. The profile is illustrated in Figure 1.18 and is the same in both axes which are transverse to the optical axis of the laser. This particular mode yields the smallest spot size and is often desired as it allows ease of focusing.

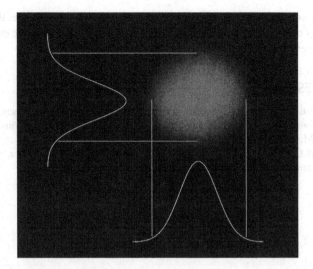

FIGURE 1.18 Gaussian (or TEM_{00}) beam shape and distribution function.

Laser cavities vary and can consist of mirrors of various types and focal lengths; however, the amplifier is essentially a waveguide and as such is subject to Maxwell's equations and the fact that a mode inside the cavity must have zero amplitude at the edges. This gives rise to modes of either cylindrical or rectangular symmetry in the output beam. Where a laser has a polarization (even a small one, imposed by optical elements inside the cavity or perhaps even the medium of the amplifier itself) rectangular symmetry is often preferred.

A common mode seen in many lasers is the TEM_{01} "doughnut" mode in which the beam takes the form of an annular ring with low intensity at the center. This mode, and several other common modes, can be seen in the photographs of actual modes in Figure 1.19. One ramification of the existence of TEM modes is that utilization of the amplifier's volume is affected by the mode that appears—some modes, naturally, have a larger "packing fraction" inside the cross section of the amplifier, which leads to higher power output.

High-order modes consume a larger area than low-order (for example, TEM_{00}) modes—in fact, this feature is often exploited to force a laser to operate in TEM_{00} mode. (Many argon and YAG lasers have small apertures within the optical cavity to force operation on this mode.) Of course, therefore, when discussing "beam area"

FIGURE 1.19 Various TEM modes.

later in this text one must consider utilization of that area—for example, the intensity of the beam is much higher in the center of a TEM_{00} beam than near the edges, which affects parameters that depend on this intensity.

REFERENCES

1. NIST: National Institute of Standards and Technology, Physical Measurements Laboratory, 2013. *Atomic Spectral Database*, http://physics.nist.gov/PhysRefData/ASD/lines_form.html
2. Weber, M.J. 2001. *Handbook of Lasers*. Boca Raton, FL: CRC Press.
3. Weast, R.C. (Editor). 1985. *CRC Handbook of Chemistry and Physics*, 66th edition. Boca Raton, FL: CRC Press.

2 Threshold Gain

Essentially every model that will be developed in this text starts with the threshold gain equation—no equation is more important in laser work and it is essential to all models that will be developed that this equation be formulated such that it accurately reflects the quality of optical elements in the laser. While the equation is, at first glance, simple enough to formulate, there are many subtleties that need to be addressed in the course of developing an accurate equation.

It is such an important concept that the majority of this chapter is dedicated to developing the equation with examples given for a number of basic laser configurations. Once developed, examples of usage of this equation in determining rudimentary laser parameters—for example, small-signal gain of a laser amplifier and minimum pump power to bring a laser to the threshold of oscillation—are examined.

2.1 GAIN AND LOSS: ACHIEVING LASING

The laser is an amplifier as well as an oscillator, and like any other oscillator, requires positive feedback to operate.

In a laser there is gain: a small optical signal traveling down the amplifier increases in amplitude with each pass; however, there are also sources of loss. An obvious loss within the cavity is the output coupler which, by necessity, removes a portion of the radiation within the cavity (intra-cavity radiation) producing an output beam. The HR represents another loss, since it is impossible to build a perfect (i.e. 100% reflecting) mirror, although this loss is generally quite small. Less obvious is absorption caused both by direct absorption of photons from the LLL to the ULL and "mechanical" losses in the amplifier including scattering of radiation by the medium.

Owing to the fact that the laser is an oscillator, total gain must equate to total losses and so if one sums all losses within a laser system together the threshold gain will be determined—this is the minimum gain required of the gain medium to overcome losses and hence initiate lasing. Where the gain delivered by the amplifying medium is less than the total losses, any useful quantity of photons is quickly absorbed and intra-cavity power cannot build.

With many lasers (especially "well-behaved" four-level systems), gain is immediate upon the onset of pumping; however, any gain developed is quickly lost to absorption and other losses in the laser (including, by necessity, the output coupler). As the rate of pumping increases, gain does as well, but only when gain exceeds total loss in the laser will a usable output beam appear. This total loss is the threshold gain of the laser system and is one of the most important parameters of a laser system (and certainly one required for essentially all models that follow).

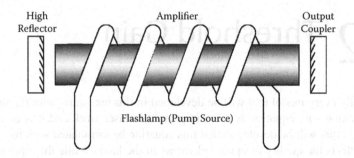

FIGURE 2.1   A simple solid-state laser.

2.2   THE GAIN THRESHOLD EQUATION

The single most important equation in laser work is the gain threshold equation. This equation allows computation of the minimum gain the laser medium must exhibit to allow the laser to oscillate. In reality, this equation represents the loss of all optical elements in the laser: absorption, cavity optics, and other (less common) intra-cavity losses.

To begin, consider the simplest case—that of a simple laser where the mirrors are fabricated directly onto the gain medium. The world's first laser, a solid-state ruby laser, was built in this form as depicted in Figure 2.1. (In the case of the first laser, the reflector coatings were deposited directly onto the ends of the polished ruby rod—they are shown here as separate elements for illustration purposes.)

In order to formulate a threshold equation for this laser, consider the losses incurred by photons during a round trip through the laser—this is best done by choosing a point inside the cavity near one cavity mirror, tracing the path of the photons (often with a pencil on a diagram of the laser), and identifying all losses and gains encountered along the way. This is done for the simple laser in Figure 2.2 in which photons start at the point marked (START). Photons move through the gain medium encountering gain (e^{gx}, as described in Chapter 1) as well as loss ($e^{-\gamma x}$) during the single pass to the left through the amplifier. Note that gain is a positive quantity while absorption (denoted γ) is negative, since a loss results in a reduction of power as per Section 1.7. Note as well that both gain and absorption are distributed across the entire length of the amplifier (denoted x in the figure) and so are measured in units of "per length" such as m^{-1} (or, for high-gain media, cm^{-1}). The photon stream then encounters cavity mirror #1 with a reflectivity of R_1—since this is the High Reflector

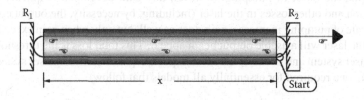

FIGURE 2.2   Round-trip path of photons through the laser.

(HR) and has essentially perfect reflectivity (values of 99.9975% are common), the numerical value is often approximated as 1.00. Photons then pass through the gain medium again (to the right) encountering both gain (e^{gx}) and loss ($e^{-\gamma x}$) a second time. Finally, the stream is reflected from the Output Coupler (OC) with a reflectivity of R_2 where a portion ($1- R_2$) passes through the mirror to become the output beam and the rest is retained as intra-cavity power, starting the cycle over again.

Mathematically, the process of formulating a threshold equation begins by multiplying all gain and loss terms and equating to unity (1) since, at equilibrium state (i.e. constant output) gain must equal loss during a round-trip through the laser. (Intuitively, if gain exceeds loss, output power will grow and if loss is larger, output will quickly be extinguished.) This is called the *unity gain equation* and is the basis for all further analysis.

The round-trip expression then becomes:

$$e^{-gx}e^{-\gamma x}R_1 e^{gx}e^{-\gamma x}R_2 = 1 \qquad (2.1)$$

$$e^{2gx}e^{-2\gamma x}R_1 R_2 = 1 \qquad (2.2)$$

The equation is then rearranged such that gain is on the left side of the equation ...

$$e^{2gx}e^{-2\gamma x} = \frac{1}{R_1 R_2} \qquad (2.3)$$

The natural logarithm is taken of both sides (and the mathematical substitution $e^a e^b = e^{a+b}$ used) to simplify ...

$$2gx - 2\gamma x = \ln\left(\frac{1}{R_1 R_2}\right) \qquad (2.4)$$

Finally, the equation is rearranged to solve for g which is now the threshold gain g_{th}:

$$g_{th} = \gamma + \frac{1}{2x}\ln\left(\frac{1}{R_1 R_2}\right) \qquad (2.5)$$

The above form of the equation is common (and found in almost every text on lasers), but when applied numerically to many real lasers it is not particularly accurate. Many lasers feature losses such as windows (necessary with many gas lasers for

In the notation used in this text, g represents a gain while the Greek character γ represents a loss.

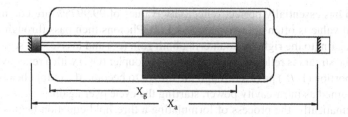

FIGURE 2.3   A gas laser with different gain and absorption regions.

vacuum integrity of the tube) and intra-cavity devices such as Q-switches (discussed in Chapter 6). Furthermore, the gain and absorption regions of a laser medium are not always equal. Still, the method by which the equation is developed is followed for all subsequent threshold gain equations in this chapter.

In many real lasers, the regions where gain and absorption occur are different and so the gain equation may be made more accurate by including two different lengths instead of simply x. Consider the cross-section of a typical helium-neon gas laser in Figure 2.3. These lasers feature a narrow central bore in which the plasma (which discharges between the small anode region on the left and the large cathode on the right) is confined and in which gain occurs. This inner bore, which has a diameter of approximately 1mm, is denoted by the length x_g (for x-gain) on the diagram. Attenuation (due to a number of factors including absorption and scattering in the medium and denoted γ) occurs in a longer region of the tube which includes the smaller gain region plus the ends of the tube where the mirrors are mounted—denoted by length x_a (for x-attenuation) on the diagram. Assuming absorption is constant throughout the length of the tube (i.e. absorption in the middle of the tube, where gain also occurs, is approximately the same as that near the ends of the tube) we can again formulate a unity gain equation in the same manner as Equation (2.1) as:

$$e^{-gx_g}e^{-\gamma x_a}R_1 e^{gx_g}e^{-\gamma x_a}R_2 = 1 \tag{2.6}$$

Rearranging, algebraically, in the same manner as the previous equation yields a new equation similar to Equation (2.5):

$$g_{th} = \frac{x_a}{x_g}\gamma + \frac{1}{2x_g}\ln\left(\frac{1}{R_1 R_2}\right) \tag{2.7}$$

Examining the preceding gain threshold Equations (2.5) and (2.7), we can see two types of losses in these equations: *point* losses which occur only at one discrete point in the round-trip of photons through the laser such as mirrors, and *distributed* losses

> Although the term g_{th} is used here for consistency (it is the most widely used) the more correct term might be γ_{th} since it really represents total loss in the laser cavity.

which occur across a larger region and are distributed as if they occur across the entire length of the gain medium. The threshold gain equation expresses all losses in the laser as a distributed loss. In other words, it computes the "gain per length" required from the amplifying medium to ensure oscillation and hence a useful output beam.

EXAMPLE 2.1   THRESHOLD GAIN OF A HENE LASER

Consider a simple HeNe laser from Figure 2.3 with different gain and absorption lengths. For these tubes, a gain threshold equation is simple to formulate since they often (but not always) lack windows or other intra-cavity elements. Mirrors are usually attached directly to the structure and are open to the narrow plasma tube inside. (Having the mirrors directly inside the vacuum envelope of the tube is possible in the HeNe laser due to the low energy of the plasma—some gas lasers have a particularly energetic plasma which prohibits such a simple arrangement.)

The gain threshold equation can hence be represented by Equation (2.7) in which there are different gain and absorption lengths. Using the following parameters,

$R_{HR} = 100\%$

$R_{OC} = 99.0\%$

$x_a = 22$cm (essentially, the length of the tube between the inside mirror surfaces)

$x_g = 18$cm (the length of the inner bore in which gain actually occurs)

$\gamma = 0.005$m^{-1} (an assumed value typical of a sealed gas laser)

the threshold gain of this configuration is determined to be:

$$+g_{th} = \frac{0.22m}{0.18m} \times 0.005m^{-1} + \frac{1}{2 \times 0.18m} \ln\left(\frac{1}{1.00 \times 0.99}\right)$$

or 0.0340m^{-1}. This is a very low value which, as required for a practical laser, is much smaller than the known gain of the medium itself (generally considered, for the HeNe laser, to be approximately 0.15m^{-1}). The criteria for a "practical laser" will be defined later in this chapter.

More complex cavity arrangements are possible, and indeed often required for certain lasers. A common arrangement, for large, high-power, lasers is a folded-cavity in which the gain medium consists of two (or more) elements. Consider a large carbon dioxide laser with an arrangement, as outlined in Figure 2.4. Two tubes, 1.5m in length each, are used as the active gain elements since a single tube 3m in length is prohibitively long (the operating voltage of a 3m tube would be excessively high, aside from which the laser would hardly fit through most doors!). Bending mirrors (R_2 and R_3) are used to complete the optical cavity.

Formulation of a threshold gain equation proceeds in the same manner as outlined for the laser in Figure 2.2, with a unity gain equation. In this case, the path of

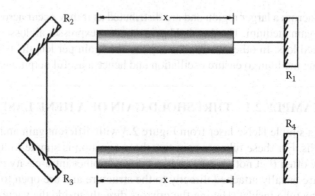

FIGURE 2.4 A "folded-cavity" arrangement.

the intra-cavity beam is traced in Figure 2.5 starting at the label marked "START" on the diagram and proceeding around the cavity in the direction indicated. The resulting unity gain equation is:

$$e^{-gx}e^{-\gamma x}R_2R_3e^{gx}e^{-\gamma x}R_4e^{-gx}e^{-\gamma x}R_3R_2e^{-gx}e^{-\gamma x}R_1 = 1 \qquad (2.8)$$

Solving Equation (2.8) for gain, the gain threshold equation for this laser becomes:

$$g_{th} = \gamma + \frac{1}{4x}\ln\left(\frac{1}{R_1R_2^2R_3^2R_4}\right) \qquad (2.9)$$

In reality, the bending mirrors are often of the same material as the HR; however, the reflectivity may be different given that these mirrors must operate at a 45-degree angle-of-incidence. The terms R_2^2 and R_3^2 recognize the fact that during a round trip through the laser, photons encounter these mirrors twice each.

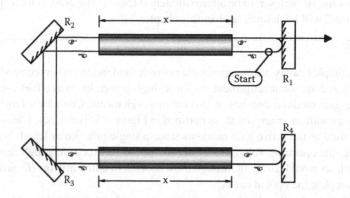

FIGURE 2.5 Optical path in the "folded-cavity" arrangement.

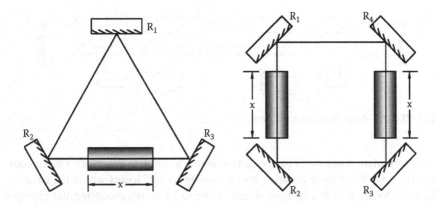

FIGURE 2.6 Ring laser cavities.

Another popular arrangement is that of a ring laser. In a ring laser, two distinct contra-rotating beams are supported in opposite rotations. One application of such an arrangement is a laser gyroscope: the difference between these two contra-rotating beams can be used to detect rotation of the laser. If unidirectional lasing is enforced, for example by including a non-reciprocal polarizer into the cavity (usually a Faraday rotator, also called an *optical diode,* which allows light to pass in only one direction), there is no standing-wave pattern within the cavity (as there is with other cavity arrangements, as described in Section 1.9) leading to unique optical properties which affect the ability of this laser to operate on a single frequency. Rings may have any number of mirrors as well as multiple gain elements, as illustrated in Figure 2.6.

The resulting gain threshold equations for these two cavity configurations shown in Figure 2.6 are:

$$g_{th} = \gamma + \frac{1}{x} \ln\left(\frac{1}{R_1 R_2 R_3} \right) \tag{2.10}$$

And, for the second arrangement with four mirrors:

$$g_{th} = \gamma + \frac{1}{2x} \ln\left(\frac{1}{R_1 R_2 R_3 R_4} \right) \tag{2.11}$$

The preceding equations contain a "distributed loss"—absorption, in this case, which is distributed across the entire length of the gain medium and measured, as gain is, in units of m^{-1}—as well as several "point" losses which occur at a single point during the round-trip transit of photons in the cavity—in this case the cavity mirrors. The reader is cautioned that, as seen in the previous example of Figure 2.4, some "point" losses occur more than once during this round trip such as windows and several intra-cavity optical devices (including Q-switches, which will be examined in Chapter 6).

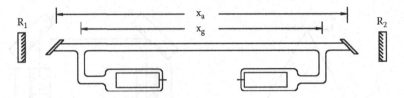

FIGURE 2.7   A gas laser with windows.

To formulate a threshold gain equation which accurately describes a more complex laser, we begin by examining how to handle "point" losses within the cavity—these are losses which are encountered at one or more points during the photon's round trip through the cavity (as opposed to the absorption term γ in the preceding case in which absorption is "spread out" over the entire length of the gain medium). In this case, incorporating these losses is a relatively simple matter of adding terms to the unity gain equation for each time these are encountered. Consider windows in a gas laser tube such as those in Figure 2.7. Once again, it is prudent to draw a line indicating a round trip through the laser to identify each time a loss is encountered leading to a round-trip unity equation as follows:

$$We^{gx_g}e^{-\gamma x_a}WR_1We^{gx_g}e^{-\gamma x_a}WR_2 = 1 \qquad (2.12)$$

where W is the "per pass" transmission of the window. Windows would be encountered four times during a round trip through the tube (once each on exiting and re-entering the laser tube). If the transmission of the window is denoted as W, Equation (2.12) would now feature the addition of the term W^4 and so solving for g_{th} yields:

$$g_{th} = \frac{x_a}{x_g}\gamma + \frac{1}{2x_g}\ln\left(\frac{1}{R_1R_2W^4}\right) \qquad (2.13)$$

One important point to note here is that W is the **portion retained in the cavity**—if a window was specified as having, for example, "2% loss per pass," then $W = 0.98$ (this is the opposite of cavity mirrors in which reflectivity is used; however, in that case, too, the portion retained in the cavity is used). Another important point is that W is rated as transmission "per pass" through the window. It is just as possible that the window might be rated on the basis of "per interface," meaning the amount of transmission where the incident radiation passes from air to window material

In the gain threshold equation losses that occur at a single point, such as mirrors, are expressed as a loss distributed across the entire length of the amplifier.

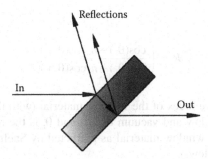

FIGURE 2.8 Reflections from an uncoated glass window.

(and again passing from the window to air). Where, for example, an uncoated glass window is used, the Fresnel equations (covered later in this section) may be used to compute the loss per surface interface (based on the index of refraction of both air and the window material). On a pass through the window, there are actually two reflections, as evident in Figure 2.8—one where the beam enters the glass (making the transition from air to glass) and a second where the beam exits the glass. The fraction of incident radiation reflected on each reflection will be equal and so the transmission on a single pass through such a window is W^2 (not $2W$, as is tempting, since the portion passing through the first interface is W and of that portion, W passes through the second interface as the radiation leaves the window so the total transmitted portion is W^2). When the "per surface" loss is known for two intra-cavity windows, Equation (2.12)/(2.13) would be modified such that W^4 term would be replaced with W^8 where W now represents the transmission "per surface interface."

In Figure 2.7, the windows are shown at an angle—Brewster's angle—which provides for essentially zero optical loss in one polarization. In a laser where windows are required (e.g. in a gas laser, for the integrity of the vacuum envelope) they may be either plane windows (at an angle perpendicular to the optical axis of the laser) or at Brewster's angle. Let's proceed to examine the reflection in each case.

If the window is uncoated, the magnitude of the reflection depends on the polarization (there are actually two reflection coefficients: one for the parallel plane and the second for the perpendicular plane) and the angle of the window to normal (this property will, in fact, be used later in the chapter to induce a variable optical loss into a laser cavity). The amount of reflection in each plane is provided by the Fresnel equations as follows:

$$R_P = \left(\frac{n\cos(\theta_i) - \cos(\theta_r)}{n\cos(\theta_i) + \cos(\theta_r)} \right)^2 \tag{2.14}$$

In the gain threshold equation, the numerical value for windows is the portion retained in the cavity.

and

$$R_s = \left(\frac{\cos(\theta_i) - n \, \cos(\theta_r)}{\cos(\theta_i) + n \, \cos(\theta_r)} \right)^2 \tag{2.15}$$

where n is the refractive index of the window material (with the interface assumed to be between the window and vacuum or air) and θ_r is the refracted angle of the incident ray inside the window material as predicted by Snell's law (included here for completeness as follows):

$$\frac{sin(\theta_R)}{1.0} = \frac{sin(\theta_i)}{n} \tag{2.16}$$

From the above equations, it can be found that for zero reflection in one polarization the window must be placed at a specific angle (Brewster's angle) within the cavity as follows:

$$tan\theta_B = \frac{n}{1.00} \tag{2.17}$$

Of course, inclusion of such a window in a laser cavity will result in a highly polarized beam, as outlined in Example 2.9 which follows later in this chapter.

The choice of window (i.e. use of a Brewster window as opposed to a plane window) depends on the specific laser employed and in many cases is dictated by the physical nature of the laser. In an ion laser, for example, uncoated windows are often used since the harsh environment inside the plasma tube (which includes highly energetic ions and extreme UV wavelengths) would quickly damage delicate optical coatings—these energetic ions will often remove such coatings in a matter of minutes. Uncoated windows are the only logical choice here and so they are placed at Brewster's angle to reduce the inserted loss of such a window to essentially zero (in one polarization, at least).

In some lasers, either type of window (plane or Brewster's) could be used. If a plane window is chosen, it must be coated with an antireflective (A/R) film to reduce inserted losses. Consider, for example, the effect of a plane, uncoated window in the cavity of a HeNe laser. Where a window is perpendicular to the optical axis, both Fresnel Equations (2.14) and (2.15) reduce to:

$$R = \left(\frac{n_1 - n_2}{n_1 + n_2} \right)^2 \tag{2.18}$$

For glass ($n = 1.5$), this would result in a reflection of 4% from each surface—a loss far too high to allow a low-gain laser to oscillate at all (and even in a high-gain laser cavity, such large losses will result in a large decrease in output power). In this case, it is clear that such windows require an A/R coating to reduce the loss due to reflections. For a single-wavelength laser, a V-coating is often preferred since it offers extremely good performance (but only at a single design wavelength).

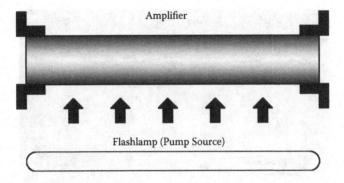

FIGURE 2.9   Cross-section of a non-uniformly pumped ruby rod.

While the A/R-coated option is not possible with many lasers (since one coated surface of the window must be open to the plasma), the relatively low energy of the plasma in a HeNe laser allows A/R coatings on the surface of the window facing the plasma without damage. In most commercial HeNe lasers, however, windows are not used at all but rather dielectric cavity mirrors (which, like A/R coatings, are also relatively delicate) are affixed directly onto the plasma tube, reducing overall inserted losses.

Finally, Brewster windows may be included inside a laser cavity solely for the purpose of polarizing the intra-cavity, and hence the output, beam. Examples of this application will be seen in this chapter as well as several chapters to follow.

In a gas laser, absorption is generally small (on the order of 0.005m^{-1} to 0.05m^{-1}), and is much lower than gain so that for many gas lasers even the simple form of the equation (as outlined in Equation 2.5, and with x representing the length of the gain element) provides a reasonable approximation; however, in a solid-state laser, absorption can be quite large (especially in a three-level or quasi-three-level laser) and so the more accurate form provided here (with separate absorption and gain lengths) must be used. To illustrate this, consider a ruby laser in which a portion of the laser rod is unpumped (the ends of the rod might, for example, be shielded by the mounting for the rod, usually using O-rings to seal the cooling water jacket around the rod and flashlamp). Such a mounting arrangement is outlined in Figure 2.9 as well as in Figure 2.10, with the former depicting the cross-section of a solid-state amplifier rod pumped by a linear flashlamp in which some of the pump radiation is occluded. In Figure 2.10 the metal cylinders shown at each end of the rod act as holders and seal the rod to the watertight housing that cools the flashlamp surrounding the rod.

While pumped regions of the rod will exhibit a large gain, unpumped regions will exhibit large loss, especially in a three-level laser. To understand the reason for this loss, consider Figure 2.11, showing the energy levels and populations of atoms (as "clouds")

Large losses in the s-polarization demonstrate why a laser employing Brewster windows will be highly polarized.

FIGURE 2.10   A mounted (bottom) and unmounted (top) ruby rod.

for both three- and four-level laser amplifiers. In an unpumped three-level laser medium most (if not essentially all) atoms reside at the ground state and so the potential to absorb an incident photon is quite large (since the rate of absorption is proportional to the number of atoms at the lower atomic level of a transition). In a ruby medium (which is three-level), the absorption coefficient can be as large as $25 m^{-1}$ (this will be calculated in Section 2.5).

In contrast, the lower-lasing level in a four-level laser medium is essentially empty (except, perhaps, for a small population of atoms pumped to the lower level by thermal means). When not pumped, most atoms reside in ground state which is not the LLL for the lasing transition. With only a small population of atoms in the LLL, the

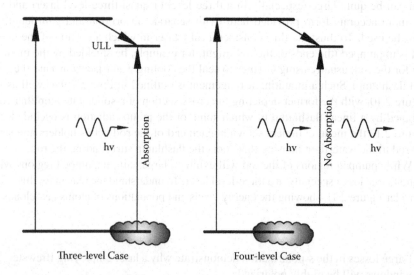

FIGURE 2.11   Absorption in unpumped three- and four-level media.

rate of absorption is very small and so a four-level medium, when unpumped, makes a poor absorber (a desirable situation).

EXAMPLE 2.2   THRESHOLD GAIN OF A NON-UNIFORMLY PUMPED RUBY LASER

For a simple ruby laser with two mirrors and an unpumped region (as per Figure 2.9) the threshold gain equation for this laser may be computed in a similar manner to Equation (2.7) by assuming x_u is the length of the rod which is unpumped and x_p the length of the rod which is pumped (where $x_u + x_p$ equals the total length of the rod) as follows:

$$e^{gx_p}e^{-\gamma x_p}e^{-\gamma_u x_u}R_1 e^{gx_p}e^{-\gamma x_p}e^{-\gamma_u x_u}R_2 = 1$$

So, for each pass through the amplifier rod, photons encounter gain in a pumped length of the rod (gx_p), attenuation in the pumped length ($-\gamma x_p$) and attenuation in the unpumped length ($-\gamma_u x_u$) where γ_u is the attenuation in the unpumped region (which, as discussed, can be very large). This equation can then be solved to yield a threshold gain equation of:

$$g_{th} = \gamma + \gamma_u \frac{x_u}{x_p} + \frac{1}{2x_p}\ln\left(\frac{1}{R_1 R_2}\right)$$

The gain medium must hence produce higher gain in the pumped region to compensate for losses in the unpumped region.
   Assume the following numerical parameters for this laser:
   R_{HR} = 100%
   R_{OC} = 90.0%
      x = 7.5cm (the length of the entire rod)
      x_u = 1cm (the length of the unpumped region)
      x_p = 6.5cm (the length of the pumped region)

Now, assuming an absorption coefficient of 2m^{-1} for pumped regions (a minimal value primarily due to scattering and attenuation in the material) and 25m^{-1} for unpumped regions (a figure which may be calculated according to the method outlined in Section 2.5 to follow), the threshold gain for this laser, numerically, is 6.7m^{-1} which is an enormously high value. Although the gain of the ruby medium can span up to 25m^{-1}, few practical lasers can achieve this gain, and a figure of 17m^{-1} is a more reasonable value (and is still quite large

---

Certain four-level materials with an LLL close to ground can exhibit three-level behavior. These *quasi-three-level* lasers are covered in Chapter 5.

for a laser). With the laser rod generating this gain, a large portion of the gain will be consumed in simply overcoming this loss, resulting in poor output. Consider now the same laser but with uniform pumping along the entire 7.5cm rod. The gain threshold formula assumes the simpler form already presented and the threshold gain calculates to 2.7m⁻¹, far lower than the case with the unpumped region of the rod. If the rod is assumed once again to exhibit a gain of 17m⁻¹, far more of this gain goes into the production of the output beam and so a far higher pulse energy results. The expected pulse energy for such a laser is calculated in Chapter 6.

EXAMPLE 2.3　HANDLING DISTRIBUTED LOSSES

As an example of the inclusion of a distributed loss, consider that some lasers exhibit multiple simultaneous output wavelengths (for example, argon ion lasers). A specific wavelength can be selected utilizing a wavelength selector consisting of a prism within the optical cavity, as shown in Figure 2.12. By rotating the prism and the HR (labeled R_2 in the figure), only one unique wavelength will be fed back into the laser for amplification. The prism, being a relatively thick material and having a finite optical path within, presents a distributed loss.

We begin by formulating a threshold gain equation for this laser. Ignoring any differences in gain and attenuation length for simplicity (and instead using a simplified x to represent the length of the gain element), the unity gain equation becomes:

$$e^{-gx}e^{-\gamma x}R_1 e^{gx}e^{-\gamma x}e^{-\gamma_{PR}x_{PR}}R_2 e^{-\gamma_{PR}x_{PR}} = 1$$

where x_{PR} is the length of the optical path inside the prism and γ_{PR} is the attenuation of the prism material. This is basically Equation (2.1) with the addition of a distributed loss within the material of the prism. The resulting threshold gain equation is:

$$g_{th} = \gamma + \gamma_{PR}\frac{x_{PR}}{x} + \frac{1}{2x}\ln\left(\frac{1}{R_1 R_2}\right)$$

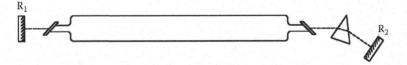

R_1

R_2

FIGURE 2.12　Wavelength selector in an ion laser.

2.3 THE TALE OF TWO GAINS: g_0 AND g_{th}

At this point it is prudent to remind the reader of the meaning of the various gain figures covered so far (since a third figure will be added in the next chapter). g_{th} is the threshold gain and is a characteristic not of the laser medium but of the optics of the laser system. In reality, g_{th} represents losses of the laser, losses that must be overcome in order for the laser to oscillate. All of the sources of these losses are optical and include absorption of the amplifier medium, losses at the cavity mirrors (a desirable loss at least for the output coupler since the "lost" portion of the intra-cavity beam becomes the output beam), and odd losses such as tube windows and intra-cavity optics. One could call it "threshold loss" since this is truly what it represents.

Assume the following numerical parameters for this laser:

$R_{HR} = 100\%$

$R_{OC} = 95.0\%$

$x = 90\text{cm}$

$\gamma = 0.05\text{m}^{-1}$

A good grade of quartz has an optical absorption of 0.020cm^{-1} and the optical path inside the prism itself is approximately 0.5cm.

Using the equation developed above, the threshold gain is found to be 0.0896m^{-1}, significantly larger than without the inclusion of the prism (so the inclusion of this intra-cavity element will have an effect on output power).

Alternately, one could express the loss from the prism as a "point loss" (i.e. a loss which occurs at a single point in the round trip through the cavity). The actual optical transmission of the prism on a single pass is then:

$$e^{-\gamma_{PR} x_{PR}} = e^{-0.020 \times 0.5} = 0.990$$

or 1% loss per pass. (This treatment is similar to the manner in which the loss of a window was expressed earlier.) In this case, the threshold gain equation becomes similar to Equation (2.13):

$$g_{th} = \gamma + \frac{1}{2x} \ln\left(\frac{1}{R_1 R_2 P^2} \right)$$

where P now represents the transmission of the prism for a single pass (i.e. 0.99). The two approaches are functionally identical, although treating a distributed loss as a point loss is an approximation since this assumes the intensity remains constant throughout the distributed loss (which it does not—it decreases during the transit through the material). Regardless, for a small loss such as this, either methodology will render a similar answer. (With a larger element, the difference between the two approaches would be more dramatic.)

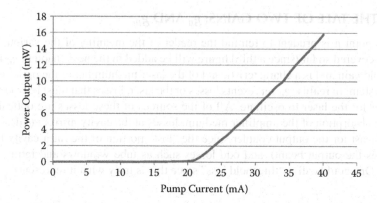

FIGURE 2.13   Diode laser output as a function of pump current.

The second gain figure introduced in this chapter, g_0, is the small-signal gain of the laser medium and represents the maximum gain which a laser amplifier can generate. (We shall see in Chapter 3 why we term this the "small-signal" gain.)

When pumping begins (at least in a four-level laser), gain begins immediately; however, the gain is insufficient to overcome losses in the laser (some of which, such as the output coupler, are unavoidable in a practical laser). As the rate of pumping increases, gain increases as well until finally reaching the threshold point where gain equals loss—this is, of course, threshold. Further increases in pumping rate increase gain which then contributes to the usable output of the laser (a concept which will be covered extensively in the next chapter).

Gain, as a function of pumping rate, can be seen in the output of a semiconductor laser as shown in Figure 2.13. As diode current increases, no significant output is seen from the laser ("significant," since output is indeed evident even with a very low pump current; however, this is primarily spontaneous emission in which the laser diode is functioning as a light emitting diode).

A number of useful parameters may now be determined from this graph; for example, the threshold of the laser is found at about 21mA—by calculating the threshold gain of the device one may now equate this pump current to that figure (i.e. the diode medium generates a gain of g_{th} at a current of 21mA).

After threshold is reached, the laser will oscillate with output increasing as a function of pumping rate (also evident in Figure 2.13). The actual gain of the amplifier stays "capped" at the threshold value, however; otherwise the output of the laser would grow indefinitely—in other words, the actual gain of the medium "burns down" to an equilibrium value of g_{th}. This process of "burning down" gain produced in the medium is called *gain saturation* and is covered in detail in Chapter 3. For the

g_{th} is a characteristic of the optical losses of a system, not the amplifier medium.

duration of this chapter, however, we examine application of the basic gain threshold equation to a variety of problems.

2.4 APPLICATION OF g_{th}: DETERMINING g_0

Unto itself, g_{th} is an important parameter that characterizes a laser (or at least the optical cavity of a laser); however, it may also be used to determine the small-signal gain (g_0) of a laser medium. In a working laser, small-signal gain must exceed g_{th}. If small-signal gain just equals the total losses of a cavity (g_{th}), the laser barely oscillates (i.e. with a very low output power). Optimally, small-signal gain must exceed threshold gain by a large margin (covered in Chapter 3).

Imagine a laser in which a variable loss is inserted into the optical cavity. Assuming the threshold gain of the cavity is less than the gain of the amplifier medium, the laser should oscillate, producing output. Now, increase that loss in the cavity, at which point output power from the laser will decrease until the laser just barely oscillates—at this point, by definition, the gain of the amplifier (denoted g_0) is equal to the sum of all losses in the cavity (i.e. g_{th}). By computing the g_{th} of the cavity at this point, g_0 is found.

The actual inserted loss can be accomplished by a variety of optical components depending on the loss required. A low gain laser, such as a HeNe laser, features an extremely low loss cavity (also called a "high Q" cavity, where Q represents the quality factor of the cavity). A typical HeNe laser features an HR with a reflectivity of over 99.99% and an OC with a reflectivity of 99% (and hence a transmission of only 1%). Factoring in all round-trip losses (including the two cavity mirrors and attenuation in the amplifier), these lasers typically have a loss of only a few percent. Consider, though, that the "typical" gain of the HeNe is $0.15m^{-1}$ so that for a 30cm laser, the gain is only around 5%. Total round-trip losses, then, including any inserted loss, must not exceed this value, so a variable loss ranging between zero and a few percent is required.

Recall from Section 2.2 that a window at Brewster's angle exhibits zero loss in one polarization and approximately 8% in the other polarization. If such a window were inserted into a laser cavity, it would most certainly polarize the beam (since 8% loss in the s-polarization will not allow the laser to oscillate); however, the laser will oscillate in the p-polarization and will be governed by the threshold equation (with zero inserted loss in that polarization). Now, if that same window is turned away from Brewster's angle such that the loss increases, the output power decreases and eventually reaches a level where the laser is barely oscillating—the intra-cavity power will be extremely low at this point, which is why g_0 is termed "small-signal gain." Knowing the angle of the window in the optical cavity (it can be measured), the loss of the window may be found using the appropriate Fresnel equation from the previous section (Equations 2.14 and 2.15) and the gain found. An arrangement of this type is seen in Figure 2.14, in which the intra-cavity beam was made visible with "fog."

Actual gain decreases as photon flux increases (as outlined in Chapter 3). Small-signal gain is hence the maximum gain an amplifier will exhibit.

EXAMPLE 2.4   DETERMINING THE GAIN OF A HENE LASER

The gain of a HeNe laser was measured in the manner depicted in Figure 2.14. This laser was found to operate at maximum output power, as expected, when the glass window (with an index of refraction $n = 1.57$) was set for Brewster's angle of 57 degrees. The window was then rotated until lasing was found to cease at 43 degrees.

The laser has the following physical parameters:

$R_{HR} = 100\%$

$R_{OC} = 99.0\%$

$x_a = 35 \text{cm}$

$x_g = 29.5 \text{cm}$

$g = 0.005 \text{m}^{-1}$

First, the gain threshold equation for the laser is developed to include the cavity optics (which, in this particular case, includes two anti-reflective coated plane windows on the tube) and the inserted loss of the window:

$$g_{th} = \gamma \left(\frac{x_a}{x_g} \right) + \frac{1}{2x_g} \ln \left(\frac{1}{R_{HR} R_{OC} W^8 G^4} \right)$$

where x_a is the length over which attenuation occurs, x_g the length over which gain occurs, W the "per surface" loss of the tube windows (each end of this particular amplifier tube is sealed with a plane window with anti-reflective coatings), and G is the "per surface" reflection from the tilted glass slide (calculated using the Fresnel equations).

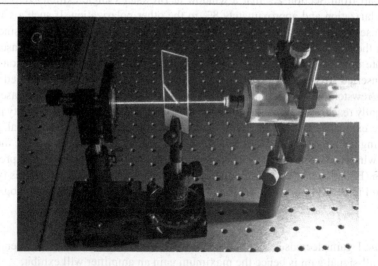

FIGURE 2.14   A HeNe laser with inserted loss.

Next, the actual loss of the window (at the point where lasing ceases) is calculated. Snell's law (Equation 2.16) is first used to calculate the refracted angle inside the glass:

$$\frac{sin\theta_R}{1.0} = \frac{\sin(43°)}{1.57}$$

The refracted angle is hence calculated to be 25.75°. Next, Fresnel's equation is used to compute the reflection loss at each glass surface:

$$R_P = \left(\frac{1.57\cos(43) - \cos(25.57)}{1.57\cos(43) + \cos(25.57)} \right)^2$$

The reflection at each surface is hence calculated to be 0.01459. The value for G in the threshold equation developed in this example is the portion retained inside the cavity (1 − 0.01459 or 0.9854).

Assuming the internal windows on the tube have a reflective loss of 0.25% per surface (so that $W = 0.9975$), the small-signal gain is computed using the equation developed in this example to be:

$$g_{th} = 0.005m^{-1} \left(\frac{35cm}{29.5cm} \right) + \frac{1}{2 \times 0.295m} \ln \left(\frac{1}{0.99 \times 0.9975^8 \times 0.9854^4} \right)$$

or 0.157 m^{-1}, which is close to the accepted value of 0.15m^{-1} for a helium-neon amplifier.

Using one slide to induce an intra-cavity loss does result in a source of experimental error, though, in that the physical position of the beam shifts as the slide is rotated. In a cavity consisting of one flat and one concave optic, this shift can misalign the optical cavity, resulting in a decrease in output power due not only to the inserted loss, but also to the cavity misalignment. Use of a single slide within a cavity consisting of plane mirrors can still cause experimental errors since the inevitable shift of the beam position can move the reflection to a region of more or less reflectivity on the same mirror (caused by inconsistency of the coating across the mirror or possibly dust and contamination on one region of the mirror). The solution is to use two contra-rotating mirrors (in a scheme outlined in *Lasers* by Siegman[1]) such that a shift of the beam in one direction caused by the first glass plate is reversed to the original position by a second glass plate. Such an arrangement is outlined in Figure 2.15.

Where two slides are used, the intra-cavity loss caused by the single slide is, of course, squared and so in Example 2.4 the transmission of the glass slide must be computed to the eighth power, not the fourth as shown in the example.

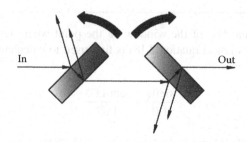

FIGURE 2.15   An inserted loss employing two contra-rotating glass slides.

Using the Fresnel equations, one may conclude that the largest practical loss that can be inserted into a cavity using a glass slide is 8%. (Larger losses are possible, at a large angle; however, experimental error will be quite large.)

High-gain lasers, such as YAGs, require a much larger inserted loss than 8% to cause lasing to cease—the inserted glass slide used in the low-gain example will simply not cause enough loss to cause the laser to stop oscillating. The simplest possible method to generate a larger loss would be to insert a series of neutral density filters directly into the cavity. Such filters, resembling grey glass discs of varying transmission, are commonly available in increments of 0.1. The transmission of such a filter is:

$$T = 10^{-OD} \tag{2.19}$$

One simply inserts a filter of low OD into the cavity, fires the laser, and if the laser still oscillates a filter of higher density is inserted. Eventually, laser oscillation ceases and at that point the total losses are summed and gain is found in the identical manner as for the low-gain laser. Of course, since filters are only available in discrete optical density values the method may not be terribly accurate; however, it will give an approximate value suitable for many analyses. For a more accurate value, a continuously variable neutral-density filter wheel (Figure 2.16) may be used in which the optical density is set by the rotational angle of the wheel. (The wheel shown in Figure 2.16 has density values imprinted around the perimeter of the wheel ranging up to 2.0 or 1% transmission.)

Figure 2.17 shows the mounting of such a filter wheel in the cavity of a YAG laser allowing gain measurement. The wheel itself can be situated anywhere in the cavity. In the arrangement shown here it is between the OC (visible in the front right) and the amplifier. The bright spot on the wheel is the beam from a HeNe alignment laser which passes through the YAG rod and identifies the location of the optical axis of this laser.

When inserting any element into the cavity of a laser, be mindful of the damage threshold of the material or coatings used.

FIGURE 2.16   A continuously variable neutral-density filter wheel.

Note that any inserted filter must be designed to handle the expected intra-cavity powers. A CW or a "long pulse" (with a pulse length over 100µs) laser often has relatively low intra-cavity powers; however, the power in a Q-switched laser (covered in Chapter 6) can be extremely large and can ablate the coatings on a filter wheel, leaving clear glass! The author had a puzzling experiment once in which the inserted loss seemed to produce the intended effect of reducing the output power as density was increased but only on the first pulse. On a subsequent pulse the laser would always return to full power. After several shots at several different density settings it was seen (on the parts of the wheel now rotated upwards) that the Q-switched intra-cavity beam

FIGURE 2.17   A filter wheel in use to determine the gain of a YAG laser.

had indeed "punched" holes in the filter coating destroying the wheel. Lesson learned: Be sure to check the damage threshold parameter of the coating (usually rated in W/cm^2) to ensure it can handle the intended power levels!

EXAMPLE 2.5   DETERMINING THE GAIN OF A YAG LASER

A variable attenuation wheel was used to determine gain of a CW diode-pumped YAG laser (in the same manner as outlined in Figure 2.17). The wheel was set for minimum transmission, the amplifier pumped at maximum diode current, and the wheel set for progressively lower densities until lasing is observed on a sensitive IR detector card placed in the output of the laser (since we are attempting to detect threshold).

Lasing was found to stop when the when transmission was less than 65%. The laser has the following parameters:

$R_{HR} = 100\%$
$R_{OC} = 90.0\%$
$x = 108$mm
$\gamma = 0.1$m^{-1}

The basic gain threshold equation, with the inclusion of the inserted loss due to the wheel (denoted as transmission of the wheel, T), is:

$$g_{th} = \gamma + \frac{1}{2x} \ln\left(\frac{1}{R_1 R_2 T^2} \right)$$

which is solved for gain as:

$$g_{th} = 0.1 m^{-1} + \frac{1}{2(0.108m)} \ln\left(\frac{1}{1.00(0.9)(0.65)^2} \right) = 4.58 m^{-1}$$

This is quite close to the "accepted" value for the gain of a CW YAG laser of 5m^{-1}.

Another alternative method of inserting a known loss into a laser cavity is the use of an intra-cavity electro-optic modulator (EOM). Often used as a Q-switch (see Chapter 6), these devices can provide a continuously variable loss where the loss is proportional to an applied voltage (although when used as a Q-switch, they are often used as a simple on/off switch where "on" is essentially 100% transmission).

Regardless of the device inside the cavity which generates the inserted loss, the methodology remains the same: insert a loss into the laser cavity until it ceases to oscillate then sum all losses—at this point gain of the amplifier medium (g_0) equals the sum of all losses (g_{th}).

2.5   AN ATOMIC VIEW OF GAIN: CROSS-SECTION

Actual gain is a function of inversion—with the reader being cautioned that inversion is the difference between the ULL and LLL populations, not simply the ULL population, since it is possible that the LLL has a significant population. In a three-level laser, for example, inversion does not begin until 50% of all atoms in the amplifier are elevated to the ULL (assuming the pump level is empty).

Gain is related to inversion by the simple relationship:

$$g_0 = \Delta N \times \sigma_0 \qquad (2.20)$$

where ΔN is the inversion and σ_0 is the cross-section (or more specifically, the stimulated emission cross-section) of the transition and can be thought of as a "target area" for an incident photon: the larger the target, the easier it is to stimulate the emission of another photon.

The relationship of Equation (2.20) requires gain in units of "per-unit length" (e.g. m^{-1}) and cross-section in units of area (e.g. m^2) so that the resulting inversion is in volumetric units. Often, ΔN must be converted into absolute units (e.g. number of atoms) in which case it may be multiplied by the volume of the active amplifier (not the device itself, but rather the volume actually involved in amplification which, in some types of lasers, may be significantly smaller).

Stimulated emission cross-sections may be calculated theoretically or may simply be researched. A great deal of research has been done for all common laser transitions of commercial value (e.g. YAG lasers) and so a quick search of physics journals and references will often yield the quickest way to obtain this figure. If it is desired to compute this parameter in a theoretical manner, it may be done by using physical constants (for example, from the NIST tables of physical data) using the following relationship:

$$\sigma(\upsilon) = Sg(\upsilon) = \frac{\lambda^2}{8\pi\tau} g(\upsilon) \qquad (2.21)$$

where $g(\nu)$ is the lineshape function which describes how gain varies as a function of frequency (i.e. how "spread out" the gain is across a range of frequencies). The exact function depends on the laser amplifier material and the method by which gain is broadened. In gases, for example, a primary mechanism which serves to broaden a transition is Doppler broadening caused by the fact that gas atoms and molecules are moving as emission occurs—this results in inhomogeneous broadening. The same (inhomogeneous broadening) is true for a solid-state laser in which the host is an irregular collection of atoms (e.g. a glass). Where a medium consists of an organized crystalline array (such as YAG), the predominant mechanism which serves to broaden transitions results is homogeneous broadening in which the shape of the gain curve remains essentially constant. This concept will be revisited in Section 3.4 where it is very significant to the way in which the gain exhibited by an amplifier saturates. Regardless of the precise mechanism involved, lineshape may be approximated depending upon the gain medium as follows:

For a homogeneously broadened medium,

$$g(v) = \frac{2}{\pi \Delta v} \tag{2.22}$$

and for an inhomogeneously broadened medium,

$$g(v) = \frac{1}{\Delta v} \tag{2.23}$$

where, in either case, Δv is the linewidth of the transition, which may be calculated or measured. Aside from the lineshape function, the lifetime of the transition is also required—this is simply the inverse of the A constant for that transition, found in a table of transitional probabilities as described in Example 1.3 in the previous chapter.

In a gas laser, the primary mechanism responsible for broadening of a transition is the Doppler effect, which is relatively easy to model. Doppler broadening occurs since gas atoms or molecules are moving at great velocities within a laser tube. Photons emitted while these particles are in motion will be shifted in frequency according to this motion in the same manner as a train whistle is observed to rise and then fall in frequency as it passes. If the particle is moving toward the observer the frequency will be shifted higher; if the particle is moving away from the observer, the frequency will be shifted lower. The actual distribution of energies of gas atoms (or molecules) follows a Maxwellian distribution and the linewidth can hence be computed by[1]

$$\Delta v = 2v_0 \sqrt{\frac{2kT \ln(2)}{Mc^2}} \tag{2.24}$$

where M is the atomic mass of the active lasing species in kg. (The above formula uses mks units: meters, kilograms, and seconds.) Knowing the atomic mass of the lasing species (in atomic mass units), this mass is computed according to:

$$M = \frac{Atomic\ Mass}{6.02 \times 10^{-23} \times 1000 g/kg} \tag{2.25}$$

Example 2.6 demonstrates the use of these formulae and the computation of cross-section for a real laser medium.

All transitions are broadened by the unavoidable effect of the lifetime of the ULL (a Fourier transform effect). Shorter ULL lifetimes lead to broader spectral widths.

EXAMPLE 2.6   CALCULATING THE CROSS-SECTION OF TRANSITIONS

Using data from NIST (see Section 1.5), the transitional probability of the 632.8nm transition (A) is found to be $0.0339*10^8$ s^{-1}. The lifetime of the level with respect to this transition (τ) is the inverse of this figure, or 295ns.

In the HeNe laser, the majority of the gas mixture is helium; however, the actual lasing transition occurs in neon and so the atomic mass of neon is used here. Assuming isotopically pure neon (Neon-20) the mass M according to Equation (2.25) is:

$$M = \frac{20}{6.02 \times 10^{-23} \times 1000 g/kg} = 3.32 \times 10^{-26} kg$$

The central frequency (v_0) is calculated as c/λ_0 (where λ_0 is 632.8nm in the case of the HeNe laser), and a moderate gas temperature of 423K is assumed which is typical for a low-energy discharge such as that in the HeNe laser. (The same is not true of ion lasers which operate at much higher temperatures.) The Doppler width of the transition is hence calculated according to Equation (2.24) as:

$$\Delta v = 2\left(\frac{3 \times 10^8}{632.8 * 10^{-9}}\right)\sqrt{\frac{2(1.38 \times 10^{23})(423)\ln(2)}{3.32 \times 10^{-26}(3 \times 10^8)^2}}$$

The linewidth is hence calculated to be 1.56GHz. This linewidth is then used in Equation (2.21) to compute the cross-section of the transition as:

$$\sigma(\upsilon) = \frac{(632.8 \times 10^{-9})^2}{8\pi(295 \times 10^{-9})}\frac{1}{1.56 \times 10^9}$$

The transition is hence found to have a cross-section of $3.46*10^{-17}$m^2.

In a similar manner, the cross-section of other visible HeNe laser transitions may be calculated, with the results summarized in Table 2.1. The cross-section of the 612nm orange transition, for example, is calculated to be $5.63*10^{-18}$m^2. Since the cross-section of the red transition is 5.9 times larger than the orange transition, the gain of that transition is 5.9 times larger. The weakest transitions of the HeNe system are the green line at 543.5nm and

TABLE 2.1

Calculated Cross-Sections for Several HeNe Transitions

Wavelength (nm)	Transition Probability (10^8 s^{-1}) (from NIST Data)	Linewidth (Hz) (Calculated)	Cross-Section (m^2) (Calculated)
543.5	0.00283	$1.82*10^9$	$1.82*10^{-18}$
594.1	0.00200	$1.66*10^9$	$1.69*10^{-18}$
604.6	0.00226	$1.63*10^9$	$2.01*10^{-18}$
612.0	0.00609	$1.61*10^9$	$5.63*10^{-18}$
629.4	0.00639	$1.57*10^9$	$6.42*10^{-18}$
632.8	0.0339	$1.56*10^9$	$3.46*10^{-17}$
635.2	0.00345	$1.55*10^9$	$3.56*10^{-18}$
640.3	0.0139	$1.54*10^9$	$1.47*10^{-17}$
730.7	0.00255	$1.35*10^9$	$4.01*10^{-18}$

the yellow line at 594.1nm which have gains fifteen times smaller than the red transition—to utilize these lines requires an incredibly high quality cavity, generally with an OC reflecting in excess of 99.5%, to oscillate.

2.6   APPLICATIONS OF THE GAIN THRESHOLD EQUATION: DESIGNING LASER OPTICS

One of the most basic applications of the gain threshold equation is the determination of the minimum reflectivity of cavity optics for oscillation (i.e. the boundary condition). With the small-signal gain known, the minimum reflectivity for a cavity optic may be computed by setting g_{th} in the threshold gain equation equal to the small-signal gain and solving for R_{OC}. In essence, we are asking "how low a reflectivity can the optics be to cause the threshold gain to be equal to the maximum gain of the amplifier?"

Assuming the laser is simple, and the threshold gain equation can be described, for example, by Equation (2.7), we may express the minimum reflectivity required by simply rearranging the equation to solve in terms of R_{OC} as follows:

$$R_{OC} = \frac{1}{e^{2g x_g} e^{-2\gamma x_a} R_{HR}} \tag{2.26}$$

Of course this value will result in the laser barely oscillating, and the output would be quite unstable since any small change in the laser will result in the laser flipping between operating and not operating. In a practical laser, the threshold gain must be well below the small-signal gain which the medium can produce: for many lasers, the rule of thumb is that the threshold gain—determined primarily by the optics employed—should be approximately one-third of the small-signal gain. An exact analytical solution is given for the optimal optical coupling in Chapter 4.

EXAMPLE 2.7   CALCULATING MINIMUM REFLECTIVITY

Consider a simple HeNe laser as described in Example 2.1 with parameters as follows:

$x_a = 22$cm

$x_g = 18$cm

$\gamma = 0.005$m^{-1}

In Example 2.1, the threshold gain was expressed by Equation (2.7) which incorporates different gain and attenuation lengths. Knowing that the small-signal gain of the HeNe medium on the 632.8nm red transition is typically 0.15m^{-1}, and that the HR is typically close to 100% reflecting, the minimum reflectivity may be determined by solving the following equation for OC reflectivity (R_{OC}):

$$0.15m^{-1} = \frac{0.22m}{0.18m} \times 0.005m^{-1} + \frac{1}{2 \times 0.18m} \ln\left(\frac{1}{1.00 \times R_{OC}} \right)$$

Yielding an answer of 0.950, the OC, then, must have a minimum reflectivity of 95% in order to oscillate. A more optimal value, however, is over 98% reflectivity. (In other words, while a mirror of 95% will allow the laser to just oscillate, a mirror with higher reflectivity will allow more intra-cavity power—and hence output power as well since this is a portion of the intra-cavity power—to build.)

Calculating cavity reflectivity goes well beyond a simple single value, though, since in many lasers the ULL is shared between many transitions. When an ULL is shared between two transitions, it is usually desirable to allow only one, desired, transition to oscillate such that the entire inversion is available for that particular transition. If broadband optics that reflect all wavelengths equally well are employed, and where one transition has considerably higher gain than another originating from the same ULL, for example, the higher gain transition often "wins," with oscillation on the weaker transition usually being extinguished completely. This is the case with a good number of popular lasers including the HeNe and the YAG lasers.

The same calculations applied already for minimum reflectivity can be used with a transition as a maximum reflectivity such that by designing an optic with a reflectivity below a desired maximum value a transition can be prevented from oscillating and so the inversion will be available to another, different, transition sharing the same ULL.

Where two transitions share the same ULL, competition exists which will lower the gain on each, or ultimately extinguish one. Wavelength selection is often used to select only one.

EXAMPLE 2.8   CALCULATING CAVITY OPTIC REFLECTIVITIES

Continuing from Example 2.7, the HeNe laser is used again. The ULL for the HeNe 632.8nm red transition is shared between many transitions in the visible and IR regions (see Section 1.5, Figure 1.11). As outlined in Section 2.5, specifically in Example 2.6, the cross-section of each transition may be computed from NIST data using Equation (2.24) to compute linewidth and Equation (2.21) to compute the actual cross-section. The gain of each transition was then computed by comparison to the known gain of the 632.8nm red transition. Table 2.2 summarizes the gain of various HeNe transitions (with commercially available wavelengths shown in boldface type).

For this example, the optics for a yellow HeNe laser will be designed—with the yellow 594.1nm transition chosen since it is the lowest gain transition of all that originate from the same ULL.

If a broadband mirror is employed, only the red transition at 632.8nm will oscillate—it will utilize essentially all of the available inversion and so the weaker lines in the orange, yellow, and green will not oscillate. In order to produce a yellow HeNe laser, optics must be carefully designed to allow lasing on the yellow line but inhibit it on any line with higher gain (including, for example, the powerful red transition as well as the aforementioned orange and green transitions).

The minimum reflectivity at 594.1nm is computed in the same manner as in the previous example with the small-signal gain figure of $0.0073m^{-1}$ now used, and so the minimum is now computed (using the methodology of Example 2.7) to be 99.96% (a very high reflectivity, especially for an OC). At all other wavelengths, the same computation is carried out; however, these reflectivities are now maximums for these "undesired" wavelengths. For example, using a small-signal gain of $0.15m^{-1}$ the reflectivity at 632.8nm was determined in Example 2.7 to be 94.95%. In that example, it was a minimum value to allow the laser to

TABLE 2.2
Computed Gain of Selected HeNe Transitions (Commercially Available Wavelengths in Bold)

Wavelength (nm)	Cross-Section (m²)	Gain (relative to 632.8nm)	Gain (m⁻¹)
543.5	**$1.82*10^{-18}$**	**0.0529**	**0.0079**
594.1	**$1.69*10^{-18}$**	**0.0488**	**0.0073**
604.6	$2.01*10^{-18}$	0.0581	0.0087
612.0	**$5.63*10^{-18}$**	**0.163**	**0.024**
629.4	$6.42*10^{-18}$	0.185	0.028
632.8	**$3.46*10^{-17}$**	1	**0.15**
635.2	$3.56*10^{-18}$	0.103	0.015
640.3	$1.47*10^{-17}$	0.425	0.064
730.7	$4.01*10^{-18}$	0.116	0.017

TABLE 2.3
Calculated Reflectivities at Various Wavelengths
for a Yellow HeNe Optic

Wavelength (nm)	Gain (m⁻¹)	Reflectivity (%)	Min/Max?
543.5	0.0079	99.93	Max
594.1	0.0073	99.96	Min
604.6	0.0087	99.91	Max
612.0	0.024	99.34	Max
629.4	0.028	99.22	Max
632.8	0.15	94.95	Max
635.2	0.015	99.66	Max
640.3	0.064	97.95	Max
730.7	0.017	99.60	Max

oscillate at that wavelength, while in this case it is a maximum to ensure it does not oscillate. Table 2.3 illustrates the calculated minimum and maximum reflectivities at each key wavelength of the HeNe laser—these would become targets for an optics design package used when fabricating these thin-film mirrors.

When graphed, potentially, the reflectivity curve of such an optic might resemble Figure 2.18 (where the points shown are the target reflectivities from Table 2.3). This illustrates how optics are designed to select a specific wavelength in a laser medium which supports many different wavelengths. Notice how "sharp" the spectral response of the mirror must be: a very high reflectivity must be achieved at 594.1nm (> 99.9%); however, the reflectivity at 632.8nm (less than 40nm away) must be less than 94.95% to ensure that this powerful transition does not oscillate (for if it did, it would certainly reduce the inversion available to the 594.1nm transition to the point where it simply would not oscillate, or at least oscillate with any appreciable output power).

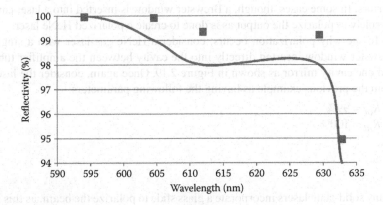

FIGURE 2.18 Reflectivity curve for the yellow optic.

Although the HeNe is used here as an example, the methodology applies equally well to gas lasers such as argon and krypton ion types (which also feature multiple transitions, some of which share levels) as well as solid-state lasers such as the YAG. In the case of the Nd:YAG laser, oscillation can occur on four main wavelengths in the near-IR: 946nm, 1064nm, 1123nm, and 1319nm, plus a few others further in the IR. The importance here is that second-harmonic generation is often employed with such lasers to yield visible output at 473nm, 532nm, 562nm, and 660nm and so selection of the fundamental frequency in the IR is important (and, like the HeNe, these transitions all share a common ULL and have different gains with 1064nm being the line with the highest gain).

Note that it was assumed that the HR has 100% reflectivity at each wavelength listed—if that optic has a different reflectivity, the product of the two must be less than, for a maximum reflectivity (or greater than, for a minimum reflectivity) the value computed. If the HR was, for example, 99.99% reflective at 594.1nm, the minimum reflectivity of the OC at that wavelength would be 99.97%. (The HR optics on a common HeNe laser are often well over 99.99% reflecting.)

The example for the HeNe shown here is incomplete since only the visible lines were covered here. In reality, the highest gain transition is not the red 632.8nm transition but instead an IR transition at 3391nm which originates at the same ULL and terminates at the 4p level of neon. This transition has a gain 44 times stronger than the red transition and so it must be suppressed either by optics with low reflectivity at this wavelength (in this example, the optic would need to have a reflectivity below 9.3% at this wavelength to prevent oscillation) or by intra-cavity absorption (which can be accomplished by borosilicate glass, which absorbs IR strongly).

EXAMPLE 2.9   POLARIZATION IN A HENE LASER

The addition of a Brewster window inside a laser cavity will polarize the output. In some lasers (such as gas ion lasers) the use of Brewster windows to seal the plasma tube is unavoidable. Unlike the common HeNe laser, mirrors cannot usually be connected directly to the plasma tube of an ion laser since highly energetic ions in the tube will rapidly damage the delicate coatings on these mirrors. In some cases, though, a Brewster window is inserted into a laser cavity solely to polarize the output as is done to create a polarized HeNe laser.

To see why polarization occurs, consider a HeNe gas laser with a single Brewster window inserted directly into the cavity between the amplifier tube and one cavity mirror as shown in Figure 2.19. Once again, consider the laser from the previous example as having the following parameters:

$R_{OC} = 99\%$

$R_{HR} = 100\%$

Many solid-state lasers incorporate a glass slide to polarize the beam, as this is required for some types of Q-switches (covered in Chapter 6).

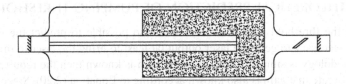

FIGURE 2.19 A polarized HeNe with internal Brewster window.

$x_a = 22$cm
$x_g = 18$cm
$\gamma = 0.005\text{m}^{-1}$

A glass slide is added inside the cavity between one mirror and the plasma tube as outlined in Figure 2.19. Many commercial HeNe lasers feature a longer metal "stem" onto which the HR is mounted specifically for this purpose—where a polarized model is desired a window is simply affixed inside this stem at the time of manufacture. This window is usually directly within the vacuum envelope for simplicity, stable mechanical design (retaining the "one piece" configuration without external mirrors), and convenience since it does not require cleaning. The gain threshold for this laser then becomes:

$$g_{th} = \gamma\left(\frac{x_a}{x_g}\right) + \frac{1}{2x_g}\ln\left(\frac{1}{R_{HR}R_{OC}T_X^2}\right)$$

where T_x is the "per pass" transmission of the glass slide in either the s- or the p-polarization (i.e. it is actually T_s or T_p). This is similar to Equation (2.13) developed previously except without the addition of tube windows. Where a single "per surface" reflection is considered from the inserted glass slide the figure would be taken to the fourth power as it was in Example 2.4. Assuming the glass slide is made of borosilicate (which absorbs 3.39μm IR well and so is well suited for this purpose—it serves to inhibit the oscillation of the powerful IR line which also shares the ULL and must hence be suppressed) and has an index of refraction of 1.57, Brewster's angle is 57.51 degrees. Snell's law then predicts a refracted angle of 32.49 degrees. Using the Fresnel equation of (2.14), loss in the p-polarization is zero, as expected, as this is the definition of Brewster's angle and so $T_p = 1$. Using Equation (2.15), reflection loss in the s-polarization is 0.178 at each air-to-glass interface, and so the "per pass" transmission (T_s) is 0.676.

Using the equation above, the threshold gain in the p-polarization is calculated to be 0.0340 m^{-1} and in the s-polarization 2.22 m^{-1}. Since the p-threshold gain is less than the known gain (g_0) of the HeNe medium, the laser will oscillate in that polarization; however, in the s-polarization the threshold gain is many times higher than the gain which the HeNe amplifier can produce, so the laser will certainly not oscillate in this polarization. Hence, inclusion of this glass slide will result in total polarization of the intra-cavity, as well as the output beam, of this laser.

2.7   A THEORETICAL PREDICTION OF PUMPING THRESHOLD

Knowing the threshold gain of a laser, it is also possible to predict the pumping intensity necessary to generate laser output (at least to bring the laser to threshold). The methodology is simple: if the threshold gain is known then the required inversion to generate that gain is also known according to Equation (2.20). Since the rate of decay from the ULL is the inverse of the lifetime of that level (i.e. the population of the level decays at a rate of $1/\tau$), the rate at which the ULL must be populated (and hence the rate at which pumping must occur) will also be known.

The rate of decay from the ULL is hence:

$$\frac{dN_{ULL}}{dt} = \frac{\Delta N}{\tau} \tag{2.27}$$

where τ is the lifetime of the ULL (as computed per the method outlined in Section 1.5). The reader is cautioned that ΔN is often calculated in volumetric units (i.e. it is actually inversion density in units of m^{-3}) and so in the above equation the resulting rate would also be in volumetric units (i.e. "per cubic meter, per second"). If an answer in "absolute units" of "number of atoms per second" is required, this figure must be multiplied by the volume of the amplifier as:

$$\frac{dN_{ULL}}{dt} = \frac{\Delta N}{\tau} V \tag{2.28}$$

Finally, substituting gain for inversion as per Equation (2.20) yields:

$$\frac{dN_{ULL}}{dt} = \frac{g_{th}}{\sigma\tau} V \tag{2.29}$$

where the result is in "absolute units." Assuming an optically pumped laser, this figure represents the rate of pump photons required—to convert this figure to actual pump power (at least assuming 100% of all pump photons result in production of lasing photons) one needs only to multiply by the energy required for each pump photon as per Figure 2.20.

As illustrated in Figure 2.20, a pump photon of larger energy than the lasing photon is required. In the case of the YAG laser, the pump photon is often (but not always) at 808nm while the lasing photon is 1064nm. This leads to the larger issue of *quantum defect* which is covered in Chapter 5.

So, having computed the rate of photons required in Equation (2.29) we need this same rate of pump photons per second so this figure is multiplied by the energy of each pump photon (in Joules) and results in an answer in J/s or watts:

$$P_{MIN} = \frac{dN_{ULL}}{dt} h\nu_{PUMP} = \frac{g_{th} h\nu_{PUMP}}{\sigma\tau} V \tag{2.30}$$

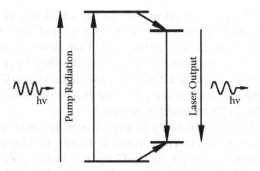

FIGURE 2.20 Pumping of the YAG system.

Of course this implies 100% efficiency of the pump source, which is never true—actual efficiency of a laser device will vary from very high (>25%, for many semiconductor lasers) to very low (<0.1%) for many gas lasers.

EXAMPLE 2.10   MINIMUM PUMP POWER OF A YAG LASER

Consider a small YAG laser with the following parameters:

$R_{HR} = 100\%$
$R_{OC} = 90\%$
Rod Size = 5mm diameter by 50mm length
Attenuation in YAG = $0.1 m^{-1}$

Assuming this is a simple laser, and that the lengths over which attenuation and gain occur are equal, Equation (2.5) may be used to compute the threshold gain for this laser to be $1.15 m^{-1}$. Given that the effective ULL lifetime of Nd:YAG is 230μs and the cross-section is $2.8 \times 10^{-23} m^2$ (both accepted values found in literature), the rate of decay from the ULL, and hence the required rate of pumping as calculated using Equation (2.29), is:

$$\frac{dN_{ULL}}{dt} = \frac{g_{th}}{\sigma \tau} V = \frac{1.15 m^{-1}}{2.8 \times 10^{-23} (230 \times 10^{-6})} (\pi \times 0.0025^2 \times 0.050) = 1.75 \times 10^{20} s^{-1}$$

Assuming a diode pumping scheme in which the YAG laser is optically pumped at 808nm (in the same manner as described in Figure 1.7, called a Diode Pumped Solid-State or DPSS laser), an 808nm photon is required for each lasing photon produced. A pump photon energy of 2.46×10^{-19} J (hv_{MP} for "minimum pump") is required and so using Equation (2.30) the required pump power is calculated to be 43.3 J/second or 43.3 Watts. This is the actual pump power absorbed by lasing ions in the rod and not simply the "raw" pump power emitted by the pump source.

In Example 2.10, the power calculated was the actual optical power delivered to the rod and absorbed by neodymium ions at the ground state. The actual pump power required now depends on the efficiency of the pump system as well as the efficiency with which pump photons are actually absorbed by the rod. The most efficient pump sources for such a YAG laser are diode lasers. Given the narrow spectral output, and one which is "tuned" to the exact absorption peak of YAG, an optical pump diode of 808nm would be required.

If a flashlamp is used, efficiency is much lower; however, peak powers are still larger than any diode currently available (hence flashlamps will continue to be an important pump source for YAG lasers). To illustrate this consider a moderate pump flashlamp of 100J with a pulse width of 1ms. The peak power of this lamp (Energy/time) is 100kW. Assuming an approximate efficiency of 1% overall (which factors in the conversion of electrical energy in the lamp to optical output, as well as the fact that the lamp emits a relatively broadband optical spectrum of which very little is actually absorbed by the YAG material), this still corresponds to 1kW of absorbed optical power—much larger than any available single-emitter diode! Knowing that the calculated threshold pump was 43.3W, a flashlamp with the efficiency described above would require only 4.3J of input to bring this laser to threshold.

The calculation of pumping threshold for a semiconductor diode laser proceeds in much the same manner as for the previous example featuring a YAG laser; however, since it is not usually optically pumped, but rather pumped by the direct injection of electrons (i.e. electrical current), Equation (2.30) (which computes the power of a stream of photons required to effect a suitable rate of pumping of the ULL) does not apply. In the case of a semiconductor laser current (in amperes) is simply the rate at which electrons are required multiplied by the charge of an electron ($1.602*10^{-19}$ Coulombs). As per Equation (2.29) we will have already computed the rate of pumping required for the ULL—in the case of a semiconductor laser this would represent the rate of electrons as per the following relationship:

$$\frac{dN_{ULL}}{dt} = \frac{J}{qt} \tag{2.31}$$

where J is the current density in units of A/m^2 and t is the thickness of the recombination (active) layer of the device. The actual geometry of a diode laser is outlined in Figure 2.21 in which the area for the current density is seen to be essentially that of the contact stripe along the top of the device: the length of the diode multiplied by the width of the active volume.

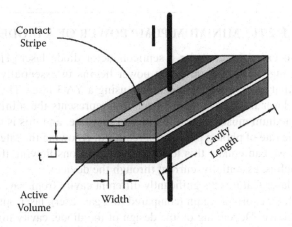

FIGURE 2.21 Structure of a simple diode laser.

One may compute J, then, from the recombination rate (i.e. dN_{ULL}/dt, as previously computed according to Equation 2.29, and in volumetric units). Finally, given

The loss of energy between the higher-energy pump photon and the laser emission photon is called quantum defect and is covered in Chapter 5.

that current through the device is distributed across the area of the contact stripe (length times width of the active region), actual device current in amperes may be computed. Alternately, for simplicity, an equation may be formulated for threshold current as follows:

$$I_{MIN} = \frac{dN_{ULL}}{dt} qV \qquad (2.32)$$

where the rate is volumetric, which yields an answer in units of amperes directly.

Semiconductor (diode) lasers are particularly efficient, and efficiencies can exceed 50% (meaning that for every two electrons injected into the device, one lasing photon is produced).

Predictions of the pump threshold current for a semiconductor diode rendered by such a model will invariably be low for a number of reasons including the fact that

EXAMPLE 2.11   MINIMUM PUMP POWER OF A DIODE LASER

As a second example, consider a semiconductor diode laser. The process of determining the minimum pump power begins in essentially the same manner as that of the preceding example using a YAG laser. The threshold gain of the laser is first computed since this represents the minimum gain which the medium must produce in order to oscillate, and this is then translated into the rate of photon production required. Knowing the rate of photon production, we can equate this to the rate of electrons flowing through the device (which is, essentially, current through the device).

A diode laser features a significantly different cavity from, say, a gas laser though: with an enormous gain (compared to a gas laser), laser optics can be far less reflective. Depending on the design of the diode, cavity mirrors may be as simple as flat, cleaved ends relying on the interface between the semiconductor material (which has a high index of refraction) and air to produce a reflection. Not all diodes use a cavity quite this simple. Many feature at least one high-reflecting coating on the rear of the diode (which allows just enough leakage to allow monitoring using a photodiode) as well as a partially reflecting coating for the OC, bringing it to an optimal reflectivity value (which can be calculated analytically as we will do in Chapter 4).

Assuming a simple laser diode with cleaved ends, the Fresnel equation of (2.18) may be used to determine the overall reflectivity of the "mirror" formed by this cleave. A diode employing GaAs (or some variation thereof, since the semiconductor material of many diodes contains a small amount of phosphorus, aluminum, or indium to modify the bandgap formed and hence the emission wavelength of the device) will have an index of refraction of approximately 3.7. Using Equation (2.18), the reflectivity of the cavity optics is hence found to be approximately 33%. This would appear to be a very low value compared to the reflectivities of optics for other lasers examined so far; however, most semiconductor lasers have an extraordinarily high gain and so can oscillate even with such low reflectivities.

To complete the calculation of threshold gain the length of the device is required. Inspecting the datasheet of Figure 2.22, one might note that this parameter is absent! It is easily found, however, by using an Optical Spectrum Analyzer (OSA) to determine the Free Spectral Range (FSR) of the device since it is, like all standing-wave lasers, an interferometer and so only certain allowed wavelengths (called longitudinal modes) are allowed to oscillate (as covered in Chapter 1).

The OSA output of the spectrum from this diode is seen in Figure 2.23 and reveals the expected modes in the output (they are especially visible when the diode is operated at just above the threshold current). This diode laser is a standing-wave laser and so the output consists of longitudinal modes—those modes being where an integer number of waves "fits" into the cavity (and where a node of zero amplitude exists at each mirror). By knowing the mode

US-Lasers: 808nm-5mW - Infrared Laser Diode

Back to Laser Diodes

INFRARED DIODE LASER DATA SHEETS   **ABSOLUTE MAXIMUM RATINGS** - (Tc=25 °C)

TECHNICAL DATA for LASER DIODE

- **Index Guided MQW Structure**
- **Wavelength: 808nm (Typ.)**
- **Optical Power: 5mW CW**
- **Threshold Current: 25mA (Typ.)**
- **Standard Package: 5.6mm**

1 laser cathode
2 common case
3 monitor diode anode

Infrared light output	808nm	**Pin Out Diagram - Style A**
Optical power output	5mW CW	
Package Type	5.6mm	
Built-in photo diode for monitoring laser output		

Items	Symbols	Values	Unit
Optical output power	Po	5	mW
Laser diode reverse voltage	VLDR	2	V
Photo diode reverse voltage	VPDR	30	V
Operating temperature	Topr	-10 ~ +40	°C
Storage temperature	Tstg	-40 ~ +85	°C

OPTICAL and ELECTRICAL CHARACTERISTICS - (Tc=25 °C)

Items	Symbols	Min.	Typ.	Max.	Unit	Test Condition
Optical output power	Po	-	5	-	mW	-
Threshold current	Ith	10	20	35	mA	-
Operating current	Iop	15	25	45	mA	Po=5mW
Operating voltage	Vop	1.9	2.1	2.5	V	Po=5mW
Lasing wavelength	8 D	800	808	820	nm	Po=5mW
Beam divergence	q F	8	11	15	deg	Po=5mW
Beam divergence	q z	20	35	45	deg	Po=5mW
Slope Efficiency (mW/mA)	0	0.1	0.3	0.6		Po=5mW
Monitor current	Im	10	100	200	m A	Po=5mW,Vr=5V
Astigmatism	As	-	11	-	m m	Po=5mW
MTTF			3000-5,000 hrs.			Po=5mW,NA=0.4
Emitter Size			1 x 4 Microns			
Emitter Distance to Cap Lens			0.3mm			
Structure			Index Guided			

FIGURE 2.22 Diode Laser Datasheet (courtesy of US-Lasers, Inc., used with permission).

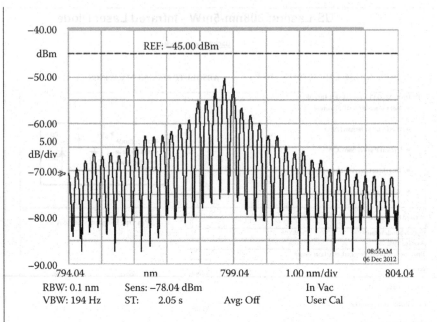

FIGURE 2.23   FSR modes in the output spectrum of a diode laser.

spacing of the optical spectrum of a device, the device length may be found according to:

$$FSR = \frac{c}{2nL}$$

where FSR is the free spectral range of the interferometer (i.e. the spacing of adjacent modes, in Hertz), and L is the spacing of the cavity optics.

An expansion of several modes of the OSA output in Figure 2.24 reveals two adjacent modes at 799.020nm and 799.318nm. Converting each to a frequency, and subtracting them, allows calculation of the FSR (the spacing of these longitudinal modes) as a frequency—in this case $1.40*10^{11}$Hz. Using Equation (1.31), and knowing the index of refraction is 3.7, the length of the device is hence calculated as 290µm.

Assuming the attenuation of the semiconductor material is 2500m^{-1}, the threshold gain of the device is now computed according to Equation (2.5) as 6323m^{-1}.

The cross-section for GaAs is approximately $1*10^{-19}$m^2 and the ULL lifetime is $1*10^{-9}$s, so the rate of recombination (dN_{ULL}/dt) in the active region is computed by Equation (2.29) to be $6.323*10^{31}$ m^{-3}s^{-1}—with volumetric units resulting since gain has units of m^{-1}, and cross-section m^2, the answer will have units of m^{-3} or "per unit volume."

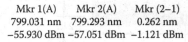

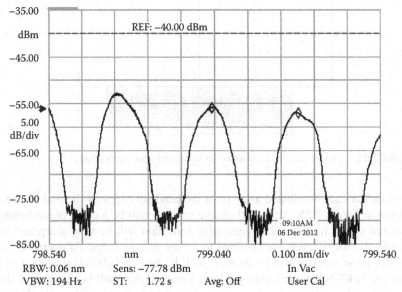

FIGURE 2.24 An expanded view of several FSR modes.

Since the rate of recombination equals the rate of pumping, electrons must be pumped into the ULL at this same rate. Now, using Equation (2.32), the current density can be calculated. From the datasheet of Figure 2.22, the thickness of the active region (t) is found to be 1μm and q is the charge of the electron ($1.602*10^{-19}$ Coulombs). The resulting current density is:

$$J = \frac{dN_{ULL}}{dt} qt = 6.323 \times 10^{31} (1.602 \times 10^{-19})(1 \times 10^{-6})$$

or $1.01*10^7 A/m^2$. Finally, multiplying the current density by the contact area for the device (i.e. length times width, or 290μm by 4μm in this case), the threshold current is predicted to be 11.8 mA.

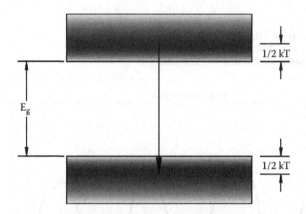

FIGURE 2.25 Charge carrier distribution in semiconductor energy bands.

charge carriers are distributed away from the bandgap edges by thermal means, as depicted in Figure 2.25. At 0K, all charge carriers will be located at the bandgap edges and so the resulting gain-generating transition will be very sharp and well defined—all energy levels are confined to a very narrow range. At operating temperatures these carriers will be located, with highest probability, at a point ½ kT above and ½ kT below the bandgap edges. These carriers will be distributed, then, according to Fermi-Dirac statistics and so gain of the laser transition will effectively be "spread out" over a wide range of wavelengths, lowering the overall gain at any one single wavelength. (The FWHM, for example, is essentially defined by the bandgap edge on one side and $E_{gap}+4kT$ on the other, since the distributions are skewed.)

Even a small flashlamp can produce power levels of 100s of kW or more so despite inefficiencies, flashlamps are still a valuable source of high powers for optical pumping of lasers.

Overall, then, gain will be decreased and accordingly the pumping current threshold of the device will increase. Experimentally, it is found that the predicted threshold current is a little more than half of the actual value measured.

REFERENCES

1. Siegman. A.E. 1986. *Lasers*. Sausalito, CA: University Science Books.
2. Csele, M. 2003. *Fundamentals of Light Sources and Lasers*. Hoboken, NJ: Wiley.

3 Gain Saturation

This chapter presents the important concept of gain saturation, perhaps the second most important basic laser concept behind the development of the threshold gain equation. Gain saturation and the mathematical description of it are key to a number of models that follow in this text.

Presented in this chapter is one of the most important models in the text, a numerical "pass-by-pass" model that demonstrates how power in a laser grows. This model, based entirely on algebra and most conveniently implemented using a spreadsheet, allows a variety of laser parameters to be predicted as well as the implementation of "what if?" scenarios. The model is flexible and may be adapted to almost any laser, and while analytic solutions that accomplish the same general task are presented in Chapter 4, this model allows the determination of a variety of parameters and helps to illustrate the method by which power grows and reaches an equilibrium state in a laser.

3.1 GAIN IS NOT CONSTANT

So far, we have examined two gain figures: g_{th}, which represents the minimum gain required from the medium to overcome losses and allow lasing to begin, and g_0, the small-signal gain of the medium which represents the maximum gain the medium can deliver.

Gain delivered by the medium starts at g_0 and "burns down," ultimately reaching an equilibrium value at g_{th}. In between these two values is the saturated gain, g_{sat}, which represents the actual gain delivered by the laser-amplifier medium for a particular value of power passing through it.

3.2 A THIRD GAIN FIGURE: SATURATED GAIN

Gain is not a constant quantity but rather varies with the photon flux inside the cavity of the laser.

Until now, we have only discussed one gain figure used to characterize a gain medium: g_0, or small-signal gain. This is the gain exhibited by the gain medium with an extremely small signal present (hence the name)—a signal so small that the ULL population remains undepleted.

To understand the concept of saturation, consider Figure 3.1, in which a small signal and a large signal both pass through a gain medium in which the pump system can only pump a constant number of atoms per second to the ULL. Assuming stimulated emission does not occur (i.e. the laser is not oscillating), the ULL population will build to a large, constant level limited by the rate at which atoms decay

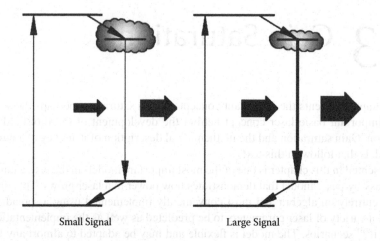

Small Signal   Large Signal

FIGURE 3.1   Reduction of gain due to photon flux.

spontaneously (when the rate of decay equals the rate of pumping into the ULL from the pump level).

Imagine, now, a small signal passing through such a system as per the left side of the diagram. Since an inversion exists (and assuming other criteria for a successful laser are met), that small input signal is amplified by the medium. In the process of amplifying this signal, the ULL will be decreased; however, since the signal is small this decrease will be negligible and so the inversion remains essentially at the maximum available, as does gain. This is the "small signal" gain we refer to and is the maximum gain that the medium will exhibit.

As the input signal increases, more ULL population is required to generate this amplification (i.e. the rate of consumption of the ULL population is large) and so the population decreases. With the decrease of the ULL population follows a decrease in gain—this is the saturated gain and represents the actual gain the medium delivers at a particular input signal.

Since inversion is not a constant value, but rather varies with the intensity of the input signal, gain is also, then, not a constant value. Practically speaking, the gain of the medium is confined to a maximum value of g_0 and a minimum value of g_{th}—it will never exceed the small-signal value since this occurs when the input signal is too small to significantly deplete the ULL population, and it will never fall below the threshold value since a gain below that value would cause lasing to cease.

With low signal intensities, the rate of stimulated emission is low. As signal increases the rate of stimulated emission will also increase.

3.3   SATURATION INTENSITY

A key parameter of a laser amplification medium is the saturation intensity (or saturation power). This is the (intra-cavity) intensity that will cause the rate of stimulated emission to equal the rate of spontaneous emission. In more practical terms, it is the intensity required to cause the gain to reduce to a value of one-half of the small-signal gain.

This parameter may be measured experimentally as well as calculated. We begin by examining the calculation. An expression for saturation intensity can be formulated from the definition for the parameter from the rudimentary rate equation analysis of Chapter 1 as follows:

$$r_{stimulated} = r_{spontaneous} \qquad (3.1)$$

Substitution is then made for the rates as follows:

$$r_{spontaneous} = \frac{N_{ULL}}{\tau_3} \qquad (3.2)$$

$$r_{stimulated} = N_{ULL}B_{32}\rho \qquad (3.3)$$

Now, knowing that Einstein's coefficient can be substituted as follows (where W is the probability of an event occurring, in this case a stimulated emission event between levels 3 and 2 hence the subscript):

$$B_{32}\rho = W_{32} = \frac{\sigma I}{h\nu} \qquad (3.4)$$

Finally, the expression for saturation intensity is:

$$I_{saturated} = \frac{h\nu}{\sigma \tau_{ULL}} \qquad (3.5)$$

Two key parameters are required for this calculation: stimulated emission cross-section (covered in Section 2.5), and the lifetime of the ULL (τ_{ULL}). It is important to realize that τ, in this case, refers to the lifetime of the lasing level—this is not simply the lifetime of the individual lasing transition but the level as a whole since there are many possible decay paths from this level which all affect the total lifetime of the level.

Intensity, in units of W/cm^2, is more correctly called irradiance (although the term *intensity* is more common and so is used here).

Radiation-trapped transitions affect ULL lifetime for the purposes of computing saturation intensities.

EXAMPLE 3.1 CALCULATING THE SATURATION POWER OF A HENE TRANSITION

We have discussed level lifetime and the method of calculating this parameter in Section 1.5; however, a slight modification is required here when discussing cross-section. Using the example of the helium-neon laser again, in Section 1.5 ten transitions were identified (see Figure 1.11) as originating from the ULL with the result being an ULL lifetime of 129ns; however, one transition was not considered: a 60.0nm transition directly between the ULL and ground which must now be considered since it indeed represents a "way out" from this level (Figure 3.2). (There is still another transition at 2416nm which is neglected since the transition probability is excessively small.)

The necessity of including this new 60nm transition is not obvious and is often overlooked. It is, in fact, radiation trapped, meaning the radiation emitted from one atomic transition is quickly absorbed by a second atom (at the same wavelength) and so, in effect, it does not affect the overall population of atoms at the ULL (as many atoms will absorb this radiation and be pumped into the $5s^1$ level as will decay from this level so it does not affect the lifetime with relevance to maintaining an inversion). Such transitions generally

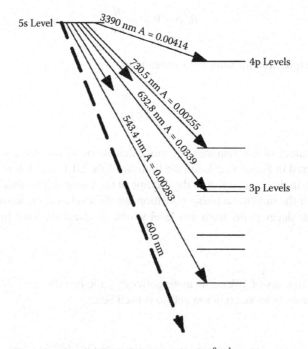

Transitional probabilities shown in units of 10^8 s^{-1}

FIGURE 3.2 Transitions originating from the neon ULL.

involve the ground state, as this one at 60.0nm does, since a LLL population is required for absorption. Without the inclusion of this transition, the lifetime was found to be 129ns. While the 60nm transition does not affect the ULL for the purposes of maintaining an inversion, it still represents a possible decay path (since we are concerned here about "the way out" of this level and not "the way in") and so it does indeed affect calculations of saturation intensity. The probability of this particular transition is very large,[1] at $0.259*10^8$ s^{-1} and so including it in the calculations yields an overall lifetime of the ULL of 29.9ns.

The existence of a pathway such as this (directly to ground state) is undesirable in a laser system and would indeed be problematic for the HeNe laser were it not for being radiation trapped. The quantum systems of many other lasers simply do not allow such a transition (due to basic quantum rules, for example where the ULL is a "D" state and the ground level an "S" state, or where spin changes during a transition, either of which results in a forbidden transition); however, in the specific case of the HeNe laser, the transition between the $5s^1$ ULL and the $2p^6$ ground level is indeed allowed.

Using the previously calculated value for cross-section for the 632.8nm transition of $3.46*10^{-17}$m^2 (from Section 2.5), and this new value for lifetime of the ULL (29.9ns), the saturation intensity of the red HeNe laser transition is calculated using Equation (3.5) to be 30.16 W/cm^2 as follows:

$$I_{saturated} = \frac{h\nu}{\sigma\tau_{ULL}} = \frac{hc}{(632.8\times10^{-9}\,m)(3.46\times10^{-17}\,m^2)(29.9\times10^{-9}\,s)}$$

To utilize this figure for calculation of power it is usually necessary to convert it into a saturation power—easily done by multiplying by the cross-sectional area of the amplifier. Assuming the physical cross-sectional area of the amplifier is completely filled with the intra-cavity beam (i.e. assuming perfect resonator alignment and a transverse mode such that the beam fills the whole area—not necessarily a realistic assumption, depending on the laser, as we will discuss later), one may find the saturation power simply by multiplying the saturation intensity by that area (πr^2).

$$P_{saturated} = I_{SATURATED} \times Area$$

For a 1mm diameter amplifier tube (in a HeNe laser, this is the bore of the central glass tube which confines the discharge), the saturation power is found to be:

$$P_{saturated} = 30.16\,W\!/\!_{cm^2} \times \pi(0.05cm)^2$$

or 236.8mW. This number, of course, depends on the beam diameter inside the amplifier, and so a bore of 0.8mm diameter yields a saturation power of 152mW.

The reader is reminded at this point that the saturation power is not a fixed limit of some type; it is simply a characterization of the system, much in the way that a speed limit is not a physical limit (for a driver may go under or over this speed)—it is simply a characteristic of the road.

3.4 SATURATED GAIN AND INTRA-CAVITY POWER

In a standing-wave laser—a configuration with two or more mirrors in which intra-cavity radiation takes the form of a standing wave with antinodes at each cavity mirror—power flows in both directions, and it is the total power which contributes to this saturation despite the fact that only power flowing in one direction, towards the OC, contributes to the output of the laser.

Consider the flow of intra-cavity power on a round trip through such a laser, as outlined in Figure 3.3. Power gain occurs as radiation (in the "reverse" direction opposite the OC, labeled I_{REV} in the figure) passes through the gain medium. This radiation is then reflected from the HR where it loses a small percentage and passes through the gain medium again in the opposite direction ("I_{FWD}"), again gaining power exponentially. Finally, cavity radiation reaches the OC where it loses a substantial percentage (that portion becoming the output beam), finally re-entering the gain medium for another pass.

The actual gain delivered by the medium is a function of the intra-cavity intensity and is described by a saturation formula, expressed first by Schulz-DuBois[2] and later used by Rigrod[3] in the development of his now-famous approach to laser modeling (outlined in the next chapter), according to:

$$g = \frac{g_0}{1 + \dfrac{I}{I_{SAT}}} \tag{3.6}$$

where I is the intensity of intra-cavity radiation and I_{SAT} is the intensity at which gain decreases to half of the original value. In a laser where radiation flows in only one direction (for example, a unidirectional ring laser), the above formula holds true; however, in a standing wave laser (as per Figure 3.3) intra-cavity radiation consists of two components traveling in opposite directions—both are responsible for the saturation of gain (since amplification of both beams depletes ULL population), but a fraction of only one beam (I_{FWD}) contributes to the output beam.

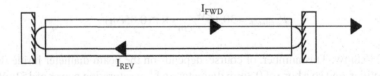

FIGURE 3.3 Flow of intra-cavity power in a standing-wave laser.

For this reason, Equation (3.6) can be approximated, for a standing-wave laser, to be:

$$g = \frac{g_0}{1 + \frac{2I}{I_{SAT}}} \qquad (3.7)$$

This approximation (called the *low-loss approximation*) assumes that the intensity of the forward and reverse beams is essentially the same. This is a reasonable assumption for a low-gain laser (such as a HeNe), and Rigrod's original paper[3] assumes that the product of the forward and reverse beams is constant. To be more correct, though, "I" must represent the average intensity on a round trip through the laser so that most correctly, $2I$ in Equation (3.7) could be substituted with "$I_{FWD} + I_{REV}$" as per Rigrod's approach—this is especially important in a high-gain laser where the intensity of the photon flux increases enormously during a single pass through the amplifier.

$$g = \frac{g_0}{1 + \frac{(I_{FWD} + I_{REV})}{I_{SAT}}} \qquad (3.8)$$

Assuming the approximation of Equation (3.7) is valid (it is for many real low-gain lasers), the actual gain delivered by the laser medium may now be calculated for any given value of incident cavity radiation. In an operating laser, gain starts at a maximum value of g_0 and decreases inversely with the value of intra-cavity radiation until finally reaching a minimum value of g_{th}. (This concept will be explored in the next section.)

At low values of cavity radiation, where $I \ll I_{SAT}$, the gain delivered by the amplifier is approximately g_0. In this region, power rises on a single pass through the amplifier according to:

$$P_{OUT} = P_{IN} e^{gx} \qquad (3.9)$$

and so an exponential increase in power is seen. As cavity intensity rises, approaching and then exceeding the saturation intensity, gain decreases rapidly, as depicted in Figure 3.4. Eventually, at large values of intra-cavity intensity (where $I > I_{SAT}$) the amplifier becomes completely saturated and power gain is now linear.

The preceding discussion assumes a medium that saturates homogeneously; however, not all amplifiers behave in this manner. As discussed in Section 2.5,

The meaning of "small-signal gain" should now be evident: the signal is so small that the amplifier is completely unsaturated and so maximum gain is exhibited.

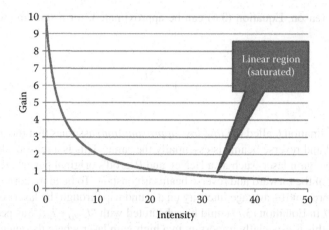

FIGURE 3.4 Gain as a function of intra-cavity intensity.

real laser transitions occur across a variety of wavelengths, resulting in a gain profile characteristic of that medium. In a medium which saturates homogeneously, the presence of laser radiation in the cavity causes a decrease in gain across the entire gain profile, meaning the gain profile retains the original shape (although gain "burns down" to g_{th}, at least at the wavelength where lasing actually occurs). In an inhomogeneously saturated medium, gain is decreased only on the actual wavelength on which the laser is oscillating, effectively burning a "hole" in the gain profile while the gain at other wavelengths (where the laser is not oscillating) is relatively large and certainly above threshold. This can result in a number of effects including the tendency of the output to "hop" between allowed longitudinal modes (often because an adjacent mode has higher gain than the one currently oscillating) and often makes single mode operation difficult to achieve. The behaviors of both types of media are outlined in Figure 3.5.

Where a medium saturates inhomogeneously, gain saturates according to:

$$g = \frac{g_0}{\sqrt{1 + \dfrac{2I}{I_{SAT}}}} \tag{3.10}$$

The general rule of thumb is this: where a material has no order or periodic structure (such as a gas or a glass), atoms behave independently and so inhomogeneous saturation occurs, whereas if the material has the form of, for example, a periodic crystal, homogeneous saturation usually occurs, since atoms in the crystal

Spectral hole burning, which occurs in inhomogeneously saturated media, changes the gain curve and can result in mode hopping.

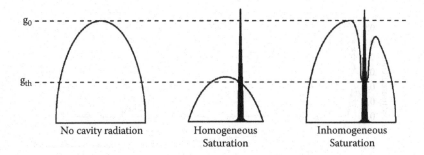

FIGURE 3.5   Homogeneous and inhomogeneous saturation.

structure are linked so that what happens to one atom in the crystal affects the crystal as a whole.

Theoretically, all transitions saturate homogeneously to some degree due to the finite ULL of the system (called *lifetime broadening*)—an effect of the Fourier transform of that lifetime—and so a short ULL lifetime results in a large broadening of the transition. There are many mechanisms which serve to broaden a transition and cause saturation of a certain type, but one mechanism usually dominates and it is based on this dominance that we classify a transition as "homogeneously" or "inhomogeneously" broadened. In a gas laser, where broadening is achieved primarily by the Doppler effect (as gas molecules move toward, or away from, the observer as they emit a photon), saturation occurs inhomogeneously (at least theoretically), and so Equation (3.10) would be the most correct saturation formula to employ.

For real lasers, though, the situation is hardly clear and many exceptions apply. As you continue through this text, you will note that many models assume homogeneous saturation in, for example, a gas laser which would normally be expected to be an inhomogeneous saturation. A homogeneous saturation model is often used with the Rigrod theory (the calculus-based approach covered in the next chapter) as a convenience since it allows an analytical solution, while use of an inhomogeneous saturation model requires numerical solutions. While done as a matter of convenience, it is not far from observed behavior, though, since it has been shown[4] that a laser operating with many longitudinal modes often "fits" a homogeneous saturation model better than an inhomogeneous model. Only where lasing occurs on a single longitudinal mode does it behave more like an inhomogeneous medium. Another complication is the consideration of the spectral "holes" burned into the gain profile in an inhomogeneous medium—at high intensities the width of a "hole" can approach the natural linewidth of the transition and so once again, the homogeneous gain saturation of Equation (3.7) applies, or at least describes the situation better than the inhomogeneous formula. Finally, where gas pressures are high, collisional broadening results in homogeneous-like behavior. There are many possible reasons for "non-conforming" behavior (meaning it doesn't do what we expect it should). For these reasons, some examples in this text describing gas lasers will assume homogeneous saturation.

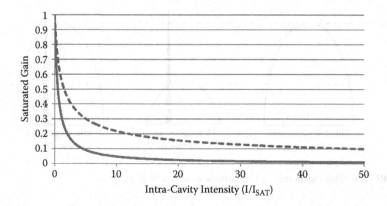

FIGURE 3.6   Saturated gain for each type of medium.

Before leaving this topic, consider that few transitions can be classified as "purely" homogeneous or inhomogeneous in behavior and so it is possible[5] to represent the saturated gain as:

$$g = \frac{g_0}{\left(1 + \dfrac{2I}{I_{SAT}}\right)^n} \tag{3.11}$$

where n is a parameter used to characterize the nature of the transition, with "1" being a purely homogeneous saturation and "0.5" being a purely inhomogeneous saturation. In reality, models have been verified experimentally showing that many gas lasers have "best fit" where n is approximately 0.8 (and once again, where n approaches 1 a purely homogeneous model is used for convenience).

Figure 3.6 outlines the saturated gain for both types of "pure" media (where "1" represents the small-signal gain). Homogeneously saturated gain media (shown as a solid line) saturate quicker than inhomogeneously saturated media (shown as a dot-ted line).

3.5 SLOPE EFFICIENCY

To recap the basic laser process: pumping is applied and gain builds until finally the gain produced by the amplifier equals total losses in the laser and an output beam appears. Gain in excess of the threshold contributes to output power. The power of the output beam, then, is proportional to the pump power applied and is

In some cases, the homogeneous profile applies best to otherwise inhomogeneous media.

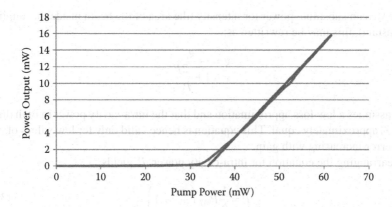

FIGURE 3.7   Slope efficiency of a laser.

characterized by the *slope efficiency* of the laser. Consider the graph of Figure 3.7 showing the output power of a laser diode as a function of pump power (both in milliwatts). Pump power is calculated from pump current knowing the bandgap voltage across the device. Until threshold is reached, output is zero, at which point output power climbs essentially linearly with increased pump power.

Slope efficiency, then, is the efficiency of the laser at converting input power to output power without consideration of threshold losses which must first be overcome. As we shall see, this is quite significant for lasers such as quasi-three-level systems (covered in Chapter 5), which may have very large thresholds; however, the laser can be designed such that slope efficiency is actually comparable to a four-level system.

3.6   PREDICTING OUTPUT POWER

A simple and useful application of this saturated gain figure is the estimation of output power. Where both the small-signal gain of the medium is known (it may be measured, as per the technique outlined in Section 2.4) as well as the threshold gain for the laser configuration (easily calculated), Equation (3.7) may be used to compute the expected power.

The basic methodology is to set g_{SAT}, the ultimate gain delivered by the amplifier, equal to g_{th}—in other words, solve the gain saturation equation in terms of the intra-cavity power required to "burn down" the gain to equal the threshold value. At threshold, gain equals loss and the laser will have reached equilibrium output power. This will predict the intra-cavity power, which must then be multiplied by the transmission of the OC to obtain the actual output power of the laser.

The process begins with the basic equation for saturated gain which, for a standing-wave laser with homogeneous broadening, is:

$$g = \frac{g_0}{1 + \dfrac{2I}{I_{SAT}}} \tag{3.12}$$

Since we can substitute power for intensity (the area of the beam inside the amplifier is constant), this may be rewritten as:

$$g = \frac{g_0}{1 + \dfrac{2P}{P_{SAT}}} \tag{3.13}$$

This assumes a low-loss approximation and that the intra-cavity power in both directions is approximately equal. This equation is hence valid only for low values of gain with error increasing with gain.

Rearranging the equation for intra-cavity power, P, yields:

$$P = \frac{1}{2} P_{SAT} \left(\frac{g_0}{g_{th}} - 1 \right) \tag{3.14}$$

and expressing this as output power from the OC (from which only a fraction, equal to $1 - R_{OC}$, exits):

$$P_{OUT} = \frac{1}{2} P_{SAT} \left(\frac{g_0}{g_{th}} - 1 \right)(1 - R_{OC}) \tag{3.15}$$

We may now substitute for threshold gain:

$$g_{th} = \frac{x_a}{x_g} \gamma + \frac{1}{2x_g} \ln\left(\frac{1}{R_1 R_2} \right) \tag{3.16}$$

$$2x_g g_{th} = 2x_a \gamma + \ln\left(\frac{1}{R_1 R_2} \right) \tag{3.17}$$

or

$$2x_g g_{th} = 2x_a \gamma - \ln(R_1) - \ln(R_2) \tag{3.18}$$

The output power can then be expressed as:

$$P_{OUT} = \frac{1}{2} P_{SAT} \left(\frac{2x_g g_0}{2\gamma x_a + \ln\left(\dfrac{1}{R_1 R_2} \right)} - 1 \right) T_{OC} \tag{3.19}$$

Of course, application to a purely inhomogeneously broadened medium will have a different saturation formula and hence a different final expression for output power.

The equations employed here use the "low loss approximation" and do not always apply for high-gain amplifiers.

EXAMPLE 3.2 PREDICTING THE OUTPUT
POWER OF A HENE LASER

Consider the same commercial HeNe laser as outlined in Example 3.1 (in which the saturation intensity of the 632.8nm HeNe transition was computed to be 30.16 W/cm² and the saturation power of a 0.8mm diameter tube was computed to be 152mW). This laser has the following cavity parameters:

R_{HR} = 100%

R_{OC} = 99.0%

x_a = 22cm (essentially, the length of the tube between the inside mirror surfaces)

x_g = 18cm (the length of the inner bore in which gain actually occurs)

γ = 0.005m⁻¹ (an assumed value typical of a sealed gas laser)

The threshold gain of this laser was computed in Example 2.1 (using Equation (2.7)) to be 0.0340m⁻¹. As per the experimental results outlined in Section 2.4, the small-signal gain of such a laser can be found to be 0.15m⁻¹ (or one can simply locate this value in published research). If one assumes that gain delivered by the amplifier ultimately "burns down" to the threshold value (g_{th}), setting the saturated gain in Equation (3.7) equal to g_{th} allows for the solution of the intra-cavity power which causes that gain decrease.

Using Equation (3.15), the output power can be found as:

$$P_{OUT} = \frac{1}{2}(152mW)\left(\frac{0.15m^{-1}}{0.0340m^{-1}} - 1 \right)(1 - 0.99) = 2.59mW$$

This is an approximation, since no compensation is made for cavity alignment or the transverse mode of the intra-cavity beam (which affects the volume of the amplifier utilized in the process); however, it does, at least, provide a ball-park figure. A commercial laser with the same parameters as those detailed in this example would have a rated output power of 4mW.

The model presented so far is simple and neglects, for example, the effects of spatial beam profile. A plane-wave, filling the entire beam area evenly, is assumed; however, in many cases the beam assumes, for example, a Gaussian profile. In such a profile, power at the center of the beam is considerably larger than might otherwise be computed and so the saturation of the gain medium in that region will be larger. The net effect can be overcome by computing a new value for saturation intensity based on the radial distribution of power in the intra-cavity beam.[6] One paper estimates that the saturation parameter of many lasers may be in error by as much as 50% due to such effects (at least in a gas laser under conditions in which inhomogeneous saturation effects apply).

Aside from mode, inspection of previous equations (e.g. Equation 3.15) reveals the dependence of saturation power on saturation intensity and beam area. A correction

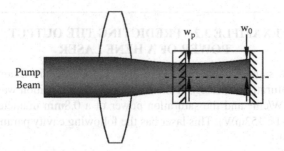

FIGURE 3.8 Beam profile in a DPSS laser medium.

is necessary for some lasers, though—for example, end-pumped DPSS lasers, in which the pump beam is often focused onto the amplifier resulting in a profile inside the amplifier as depicted in Figure 3.8, which shows the pump beam (coming from the left) focused onto a solid-state amplifier with integral cavity mirrors. In this case, the waist of the pump beam is often a different size than the intra-cavity mode. The average beam area, then, can be expressed in this laser as:

$$A = \pi r^2 = \frac{1}{2}\pi(w_0^2 + w_p^2)$$ (3.20)

where w_0 and w_p are the radii of the cavity mode and the pump beam, respectively. These radii are defined as the point where the beam intensity falls to $1/e$ (assuming a Gaussian beam profile in each case).

3.7 MINIMUM PUMP POWER REVISITED

In Chapter 2, we considered the minimum pump power required to bring a laser to threshold as:

$$P_{MIN} = \frac{dN_{ULL}}{dt}h\nu_{PUMP} = \frac{g_{th}h\nu_{PUMP}}{\sigma\tau}V$$ (3.21)

One might now recognize that if there is no significant difference between pump photon energy and lasing photon energy (what we will later, in Chapter 5, define as a small quantum defect) one may substitute for saturation in the above equation to yield:

$$P_{MIN} = I_{SAT}g_{th}V$$ (3.22)

and Volume may be expanded as Area * Length of the amplifier, with length represented by x as we have done in Chapter 2. Now g_{th} in the above equation is expressed in units of m^{-1}; however, this may be expanded to express gain in terms of a round trip (where gain now becomes a dimensionless quantity) as:

$$P_{MIN} = \frac{1}{2}I_{SAT}A(2g_{th}x)$$ (3.23)

Threshold gain may now be expanded to:

$$P_{MIN} = \frac{1}{2} I_{SAT} A(2x) \left(\gamma + \frac{1}{2x} \ln\left(\frac{1}{R_1 R_2} \right) \right) \tag{3.24}$$

and simplified to:

$$P_{MIN} = \frac{1}{2} I_{SAT} A \left(2x\gamma + \ln\left(\frac{1}{R_1 R_2} \right) \right) \tag{3.25}$$

This assumes (for an optically pumped laser, at least) that the energy of the pump photon is close to that of the lasing photon energy. Following the concept outlined in Section 2.7 (and illustrated in Figure 2.20), the pump photon energy may differ significantly from laser photon energy (i.e. a large quantum defect exists, as defined in Chapter 5), and one must multiply this minimum pump power by a factor of:

$$\frac{h\nu_{PUMP}}{h\nu_{LASER}} \tag{3.26}$$

Be forewarned, though, that there are alternative mathematical definitions for "quantum defect."

One must, of course, compensate for other efficiencies, including *quantum efficiency* (which is not the same as quantum defect but rather is an expression of the fraction of pump photons that result in production of a photon of laser radiation) since no laser is 100% efficient (for example, semiconductor diode lasers can commonly reach an efficiency of 50% and so the minimum pump power must be multiplied by two to obtain a reasonable estimate). Other lasers have efficiencies much lower than this due to a variety of factors—for example, absorption of actual pump radiation from a broadband pump source like a flashlamp, as discussed in Section 1.2.

3.8   ALTERNATIVE NOTATIONS

Depending on the source of literature, the output power and threshold pump power may be described in a simplified form which, for the sake of completeness, is developed here. The basic methodology of applying these simplifications will be used further in this text as well.

In the case of output power calculations, we begin with Equation (3.14) and expand for g_{th}:

$$P_{OUT} = \frac{1}{2} P_{SAT} (1 - R_{OC}) \left(\frac{g_0}{\gamma + \frac{1}{2L} \ln\left(\frac{1}{R1 R2} \right)} - 1 \right) \tag{3.27}$$

Multiplying both numerator and denominator by $2L$ (where L is the length of the gain element, equivalent to x_g) yields:

$$P_{OUT} = \frac{1}{2} P_{SAT} (1 - R_{OC}) \left(\frac{2g_0 L}{2\gamma L + \ln\left(\frac{1}{R1R2}\right)} - 1 \right) \tag{3.28}$$

Finally, assume a perfect HR (i.e. $R_2 = 1.00$), and define a few simplified quantities including the total attenuation of the medium:

$$a = 2\gamma L \tag{3.29}$$

in which a is a dimensionless quantity representing the sum of all cavity losses (including absorption in the gain medium, scattering, etc.) during a round trip through the amplifier. It should be noted as well that some authors use the term δ to denote this quantity instead of a—the distinction being that δ, or a, represents a dimensionless quantity while γ (used so far in this text to denote attenuation in the amplification medium) represents a "per unit length" quantity in units of m^{-1} or cm^{-1}. Simplifications such as this will be applied in later chapters for consistency with key research papers.

A further simplification can be applied to Equation (3.28) as follows:

$$ln\left(\frac{1}{R}\right) = -\ln(R) \approx T \tag{3.30}$$

where T is the transmission of the OC and so is equal to $(1 - R)$. This simplification is valid only for low values of OC transmission (<10%, which is normal for a low-gain laser such as most gas lasers). Finally, Equation (3.28) simplifies to:

$$P_{OUT} = \frac{1}{2} P_{SAT} \left(\frac{2x_g g_0}{2\gamma x_a + T_{OC}} - 1 \right) T_{OC} \tag{3.31}$$

Then, making a few substitutions to simplify the equation,

$$P_{OUT} = \frac{1}{2} P_{SAT} T \left(\frac{2g_0 L}{a + T} - 1 \right) \tag{3.32}$$

> The notation introduced here expresses all losses as dimensionless quantities (i.e. "per round trip") instead of "per unit length."

> The simplification $ln\left(\frac{1}{R}\right) = -\ln(R) \approx T$ is often applied to equations for gain threshold and laser output power.

$2g_0L$ is the total round-trip gain and is also a dimensionless quantity. (Occasionally, this will be written as g_0L where it is understood that L is the total gain length which, for a round trip, is $2L$, or even as g_0 where this is, ambiguously, equal to $2g_0L$.) One may also substitute for $P_{SAT} = I_{SAT}*A$. It is seen, then, that this equation (and perhaps other minor variants that involve other simplifications) and the previously derived solution are equivalent.

These same simplifications are often made to formulae for the calculation of minimum pump power threshold to simplify those equations as well. Starting with Equation (3.25), the same substitutions for a and T may be made to yield the simplified form that is often quoted in literature:

$$P_{MIN} = \frac{1}{2} I_{SAT} A (a + T) \tag{3.33}$$

3.9   A MODEL FOR POWER DEVELOPMENT IN A LASER

One of the most powerful (and yet simplest) models in this entire book, the "pass-by-pass" model presented here, predicts the power inside the laser cavity as a function of "pass number" through the amplifier and is entirely algebra-based. Easy to implement on a spreadsheet, the model predicts a number of laser parameters and can be used to determine parameters such as optimal output coupling, a parameter normally computed using rigorous calculus methods covered in the next chapter.

The simplest illustration of the model is provided by the development of power within a low-gain laser such as a HeNe laser. In this simplest version of the model, outlined here, each line of the spreadsheet represents one round-trip pass through the laser amplifier where the photon stream passes through the amplifier, reflecting from a perfect lossless HR (illustrated in the figure as an "empty" cloud), through the amplifier again in the reverse direction, and reflecting from the OC. This round-trip path is illustrated in Figure 3.9, where P_{INPUT1} is the initial power which then passes through an amplifier of effective length $2x$ (since the HR is 100% reflecting) and exits the amplifier with a power of $P_{OUTPUT1}$. A portion of this output passes through

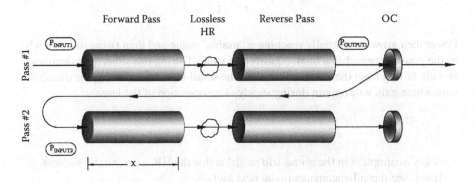

FIGURE 3.9   Power development during a round trip through the laser.

the OC to become the output beam and the portion reflected from the OC returns for the next pass through amplifier (P_{INPUT2}) with the cycle repeating many times.

To begin, the saturated gain is calculated for the entire round trip through the laser, based on the power entering the amplifier, as follows:

$$g_{PASS1} = \frac{g_0}{1 + \dfrac{2P_{INPUT1}}{P_{SAT}}} \tag{3.34}$$

The output power after a round trip through the amplifier is then computed according to Equation (3.19) which is rearranged to express gain and attenuation as a "net" gain as:

$$P_{OUTPUT1} = P_{INPUT1} e^{(g_{PASS1}-\gamma)2x} \tag{3.35}$$

Note that the effective amplifier length is $2x$ in this case, where x is the length of the amplifier. The input for the second pass is then simply the portion which reflects from the OC back into the amplifier:

$$P_{INPUT2} = P_{OUTPUT1} \times R_{OC} \tag{3.36}$$

and the output beam is comprised of the portion transmitted by the OC:

$$P_{OC} = P_{OUTPUT1} \times (1 - R_{OC}) \tag{3.37}$$

The process is, of course, repeated many times until an equilibrium output power is reached. The power starts, on the first pass, at a very low value: a good estimate would be the effective power of a single "seed" photon which originates in the laser building to a stream of photons as operation progresses. This is simply the energy of a single photon (in Joules) divided by the lifetime of the ULL as follows:

$$P = \frac{E}{\tau} = \frac{h\upsilon}{\tau} \tag{3.38}$$

Power then grows, eventually reaching a "usable" value and then rising more slowly as the power approaches a final, equilibrium value. During this time, gain decreases as well, falling from the initial value of g_0, the small-signal gain, to g_{th}, the threshold gain where gain will remain during steady-state operation of the laser.

A key assumption in the round-trip model is that the HR is essentially lossless. If not, see the enhancements in the next section.

Consider, now, an example numerical application of the preceding formulae (which will eventually become a single line from a spreadsheet modeling a laser) at a point where the input power is only 1mW—arbitrarily chosen here for this illustration. A host of parameters are required for any model, and in this case a HeNe laser is used in which assumed parameters include the small-signal gain of $0.15m^{-1}$, and a saturation power of 152mW (calculated previously based on a bore diameter of 0.8mm as per Example 3.1). The saturated gain for this particular round trip is calculated according to Equation (3.13) (with $2P$ used to approximate the intra-cavity power since, as a low-gain laser, the power increase between two subsequent passes is very small), is computed as:

$$g = \frac{g_0}{1 + \dfrac{2I}{I_{SAT}}} = \frac{0.15m^{-1}}{1 + \dfrac{2(1.0mW)}{152mW}} = 0.148m^{-1} \tag{3.39}$$

The power after a round-trip pass, then, is computed according to Equation (3.35) as:

$$P_{OUT} = 1mWe^{(0.148-0.005)(2)(0.22m)} = 1.065mW \tag{3.40}$$

The actual net gain for this round-trip pass is the saturated gain minus the attenuation (assumed to be $0.005m^{-1}$ in this case), $0.143m^{-1}$, and is the effective gain available for amplification.

Finally, a small portion of the power produced after the round trip (0.01 times 1.065mW or 10.65μW, assuming a 1% transmitting OC) exits through the output coupler to become the output beam, and the majority (0.99 times 1.065mW or 1.054mW as calculated by 3.36) becomes the starting power for the next trip through the amplifier.

While the model can be implemented using any computer programming language, perhaps the most easily available tool is a spreadsheet, which requires no actual programming. A spreadsheet may be set up such that each individual row represents a round-trip pass through the laser, allowing one to easily visualize how power increases and gain decreases as the laser operates. Copying equations from one line to a subsequent line results in the incrementing of row references so that each row will represent a subsequent pass through the amplifier.

The equations for one line are simple and follow the basic equations outlined: calculate the saturated gain based on the input power to the amplifier, compute the resulting power after a round trip (two passes) through the amplifier, remove the portion which passes through the OC (which becomes the output beam), and take the remaining (reflected) portion to be the input power for the next pass.

Parameters required for the calculation of saturated gain include the small-signal gain and saturation power. Saturation power is pre-calculated in the same manner as Example 3.1 (and is the saturation intensity multiplied by the area of the amplifier). For calculation of power after the round trip, optics parameters including attenuation and mirror reflectivity are required. All of these parameters are anchored at the top of the spreadsheet for global reference (this also allows "what if" scenarios to be

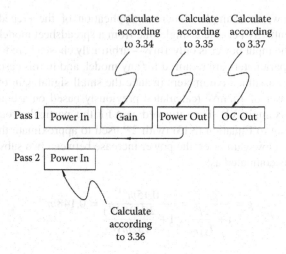

FIGURE 3.10 Spreadsheet formulae for the model.

investigated). Figure 3.10 details the layout of the spreadsheet and the formulae that are used in the model to calculate each cell.

EXAMPLE 3.3   MODELING POWER BUILDUP IN A HENE LASER

To illustrate the implementation of the model, consider a simple model for a HeNe laser using a spreadsheet to predict power after each round trip.

Examining Figure 3.11, we see the physical parameters of the model are located in cells C4 to C8. By placing common parameters at the top, they may be changed easily allowing, for example, determination of how changes to the output coupler affect the ultimate output power of the laser. (This is a useful feature of the model and allows, for example, determination of optimal OC transmission.) These cells are assigned, in this example, parameters as follows:

C4: Small-signal gain (g_0) in units of m^{-1}
C5: Saturation Power (P_{SAT}) in mW
C6: Attenuation (γ) in units of m^{-1}
C7: Length of the gain medium in units of m^{-1}
C8: Loss of the optics (transmission of the OC)

Values typical for a HeNe are shown including saturation power (calculated in a previous example).

Avoid the use of "hard-coded" constants in spreadsheet formulae: anchor parameters at the top to allow "what if" scenarios to be investigated.

	A	B	C	D	E	F
1	**Simulation: HeNe Laser**					
2						
3	**Laser Parameters:**					
4		Gain (go) =	0.15	m⁻¹		
5		Psat =	152	mW		
6		Atten (gamma) =	0.005	m⁻¹		
7		Length =	2.20E-01	m		
8		Loss (optics)	0.01	per round trip		
9						
10	Pass#	Pin	Gain	Actual Gain	Pout(amp)	Pout
11		(mW)	(m⁻¹)	(m⁻¹)	(mW)	(mW)
12	0	2.44E-12	1.50E-01	0.145	2.60E-12	2.6E-14
13	1	2.56957E-12	1.50E-01	0.145	2.74E-12	2.74E-14
14	2	2.71146E-12	1.50E-01	0.145	2.89E-12	2.89E-14
15	3	2.86119E-12	1.50E-01	0.145	3.05E-12	3.05E-14
16	4	3.01918E-12	1.50E-01	0.145	3.22E-12	3.22E-14
17	5	3.1859E-12	1.50E-01	0.145	3.40E-12	3.4E-14
18	6	3.36183E-12	1.50E-01	0.145	3.58E-12	3.58E-14
19	7	3.54747E-12	1.50E-01	0.145	3.78E-12	3.78E-14
20	8	3.74336E-12	1.50E-01	0.145	3.99E-12	3.99E-14

FIGURE 3.11 A model for the HeNe laser.

Proceeding to examine a row from the model, the first row (12) begins with a "seed" power in cell B12. An arbitrarily small power could be used here (the power must be far smaller than the saturation power); however, the most correct approach would be to use the power of a single photon (E_{photon}/τ) which assumes that the entire optical flux grows from a single "seed" photon according to Equation (3.38).

Saturated gain is then computed in the next column, C12, according to Equation (3.13), which refers to g_0 and P_{sat}, both of which are anchored at the top of the spreadsheet. By using references to cells \$C\$4 and \$C\$5, these fixed cell references will be maintained when the formula is copied to the next row (since all rows have identical formulae). The formula for saturated gain (in C12) is hence " = \$C\$4/(1+2*B12/\$C\$5)" where C4 refers to the small signal gain and C5 the saturation power.

The fixed attenuation (also anchored at the top of the sheet in cell C6) is subtracted from the computed saturated gain to yield net gain, which is computed in D12 as " = C12-\$C\$6" (this could have been done directly in the next step in which power is computed, but it is illustrated here as two discrete steps for clarity). The output power from the round trip through the amplifier (two passes, including the lossless reflection from the HR) is calculated in cell E12 using Equation (3.40) implemented in the spreadsheet as " = +B12*EXP(2*D12*$C$7)" where cell B12 is the input power for this specific round trip, D12 the effective gain, and C7 the length of the amplifier medium. The multiplier 2 denotes that two passes through the amplifier comprise a complete round trip. Finally, the portion transmitted through the OC (the output power observed) is calculated in cell F12 by multiplying by $C$8 where this cell is the transmission of the cavity optics (which represents a loss). The output from the model is plotted in Figure 3.12.

To complete the spreadsheet, the portion retained in the cavity is calculated by multiplying by (1-$C$8) and this value used as the starting power for the next round-trip (i.e. in cell B13).

All formulae are copied to subsequent rows in the sheet. The spreadsheet program automatically increments the row number when a formula is copied so that the formula " =+B12*EXP(2*D12*$C$7)" from row 12, when copied to the cell immediately below it in row 13, becomes " = +B13*EXP(2*D13*$C$7)."

A summary of all cells in the first row:

B12: Starting power

C12: Saturated Gain for this trip, = $C$4/(1+2*B12/$C$5)

D12: Actual Gain (Saturated-attenuation), = C12-$C$6

E12: Output power after round-trip through the amplifier, = +B12*
     EXP (2*D12*$C$7)

F12: Power passing through the OC, = $C$8

B13: Starting power of the next pass, = E12*(1-$C$8)

To view the output from the model, the output power and gain are graphed as a function of pass number in Figure 3.12 showing, as expected, gain (represented by the thick line) decreasing from an initial value of g_0 to a final value of g_{th} (0.0279m^{-1}) as power (represented by the thinner line) builds to the final equilibrium value. For comparison, g_{th} is calculated using Equation (2.5) to be

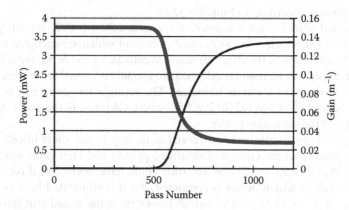

FIGURE 3.12   Output of the model.

	A	B	C	D	E	F
1	Simulation: HeNe Laser (Improved Model)					
2						
3	Laser Parameters:					
4		Gain (go) =	0.15	m⁻¹		
5		Psat =	152	mW		
6		Atten (gamma) =	0.005	m⁻¹		
7		Gain Length =	1.80E-01	m		
8		Atten Length =	2.20E-01	m		
9		Loss (optics) =	0.01	per round trip		
10						
11	Pass#	Pin	Gain	Actual Gain	Pout(amp)	Pout
12		(mW)	(m⁻¹)	(m⁻¹)	(mW)	(mW)
13	0	2.44E-12	1.50E-01	0.145	2.56E-12	2.56456E-14
14	1	2.53891E-12	1.50E-01	0.145	2.67E-12	2.6739E14
15	2	2.64716E-12	1.50E-01	0.145	2.79E-12	2.78789E-14
16	3	2.76002E-12	1.50E-01	0.145	2.91E-12	2.90675E-14
17	4	2.87768E-12	1.50E-01	0.145	3.03E-12	3.03068E-14
18	5	3.00037E-12	1.50E-01	0.145	3.16E-12	3.15988E-14
19	6	3.12829E-12`	1.50E-01	0.145	3.29E-12	3.2946E-14
20	7	3.26166E-12	1.50E-01	0.145	3.44E-12	3.43506E-14
21	8	3.40071E-12	1.50E-01	0.145	3.58E-12	3.58151E-14

FIGURE 3.13 Anchored spreadsheet parameters for improved accuracy.

0.278m^{-1}. At the same time as gain decreases, power increases from zero to the maximum equilibrium power of 3.37mW.

Compare the final value of output power to that computed using previous methods. In Example 3.2 the output power was calculated to be 2.59mW; however, the threshold gain g_{th} was computed in that example according to Equation (2.7) in which the gain and attenuation lengths are different, for a more accurate result. To make this model more accurate, then, we should consider this as well and so a few modifications are in order to the current model.

Since gain and attenuation lengths will be considered separately, they are anchored as parameters at the top of the spreadsheet, as seen in Figure 3.13.

Parameters are calculated in the same manner as previously; however, with separate lengths the formula for power after a single round-trip pass through the amplifier becomes:

$$P_{OUT} = P_{IN} e^{2 g_{SAT} x_g} e^{-2 \gamma x_a}$$

and so, expressed as a spreadsheet formula (in the cells of column E) " = +B13*EXP(2*C13*$C$7)*EXP(-2*$C$6*$C$8)" where C7 is length x_g and C8 is length x_a. Once again, parameters are not "hard coded" into formulae but rather anchored at the top allowing "what if?" scenarios to be investigated. The output of the model is seen in Figure 3.14.

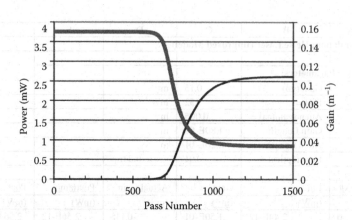

FIGURE 3.14   Output of the upgraded model.

The output is now predicted to be 2.62mW in much better agreement to the previous model of Section 3.2. The final gain is also the same as that predicted by Equation (2.7) at 0.0340m⁻¹.

The spreadsheet model is available for download from the publisher's website at http://www.crcpress.com/product/isbn/9781466582507

To verify the validity of the model, consider the assumptions made in the model. One major assumption is the use of twice the input power to the amplifier when calculating saturated gain for a round trip through the amplifier. This power directly affects gain, and a quick check of the model will reveal how much gain actually changes between two subsequent passes. A "reasonable" approximation of the model will dictate that the gain change by no more than a few percent. When the change in gain exceeds this figure, an improved approach (outlined in the next section) may be warranted.

3.10   IMPROVING THE MODEL FOR USE WITH HIGH-GAIN LASERS

The previous model, which calculates the gain and power for each round trip through the laser, makes a number of assumptions: for example, that the power between any two subsequent passes does not change appreciably (i.e. the gain is quite low and so this is a reasonable assumption, backed up by the numerical results which show no appreciable increase between passes) as well as inclusion of a lossless HR such that all of the power exiting the amplifier on the forward pass enters the amplifier on the reverse path. Both are entirely reasonable assumptions for a low-gain laser (such as the HeNe used in Example 3.3); however, when employed with a high-gain laser the model shows significant problems.

Two improvements to the model are hence (1) to compute the gain for each pass through the amplifier, both forward and reverse, and (2) compensate for any losses

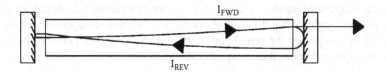

FIGURE 3.15 Beam growth in a standing-wave laser.

from the HR, since a real HR will have losses (however small) which may affect the accuracy of the model. Both assumptions reflect the physical situation in a laser as outlined in Figure 3.15, which shows the relative intra-cavity beam powers (in the vertical axis) in a standing-wave laser.

Power of the forward beam grows as the beam moves towards the right until it reaches the OC, at which point a portion is extracted to become the output beam and a portion (smaller than the incident power of the forward beam) is reflected back into the amplifier to become the reverse beam. The power of the reverse beam grows as well until it is reflected from the HR (and in the process incurring a small loss, since the HR is never truly lossless). The traversal of the beam through the laser is shown in a more logical manner in Figure 3.16. This improved model builds on the previous round-trip model, although the power of the intra-cavity beam is computed after each individual pass through the amplifier.

To begin, the saturated gain is calculated for the forward pass through the laser, based on the power entering the amplifier in the forward direction, as follows:

$$g_{FWD1} = \frac{g_0}{1 + \frac{2P_{INPUT1}}{P_{SAT}}} \quad (3.41)$$

The power increase in the forward direction is then computed as:

$$P_{OUTPUT1F} = P_{INPUT1}e^{(g_{FWD1} - \gamma)x} \quad (3.42)$$

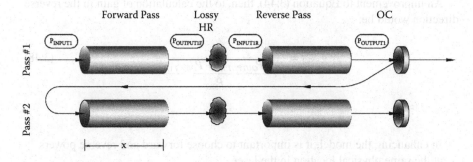

FIGURE 3.16 Single-pass enhancement of the model.

where the amplifier length x is used since we are computing this figure for only a single pass through the amplifier in one direction. We now compensate for the loss at the HR by computing the actual power entering the amplifier, because the reverse pass is the portion kept in the cavity after the reflection:

$$P_{INPUTR1} = P_{OUTPUT1F} \times R_{HR} \tag{3.43}$$

The gain is now recalculated for the reverse pass since the power is now presumably higher and so the gain will be lower:

$$g_{REV1} = \frac{g_0}{1 + \dfrac{2P_{INPUT1R}}{P_{SAT}}} \tag{3.44}$$

The power increase in the reverse direction is then computed using the updated gain:

$$P_{OUTPUT1} = P_{INPUT1R} e^{(g_{REV1} - \gamma)x} \tag{3.45}$$

As per the previous model, the input for the second pass is then simply the portion which reflects from the OC back into the amplifier:

$$P_{INPUT2} = P_{OUTPUT1} \times R_{OC} \tag{3.46}$$

and the output power from the OC is:

$$P_{BEAM1} = P_{OUTPUT1} \times (1 - R_{OC}) \tag{3.47}$$

Now, one could argue that depending on the gain of the laser, the choice of power for calculation of the saturated gain is a potential source of error regardless. Examining Equation (3.41) it is evident that the use of twice the input power is a simplification, and the more correct treatment would be to use the power in both the forward and reverse directions as per Equation (3.8). The same argument can be used with the HR. In the case of a "low-loss" HR, Equation (3.44) is still valid; however, where the HR has a significant loss that approach might lead to significant errors. Figure 3.17 outlines the forward and reverse powers that affect the gain on the reverse pass.

An improvement to Equation (3.44), then, to the calculation of gain in the reverse direction would be:

$$g_{REV1} = \frac{g_0}{1 + \dfrac{(P_{OUTPUT1F} + P_{INPUT1R})}{P_{SAT}}} \tag{3.48}$$

In enhancing the model, it is important to choose forward and reverse powers at the same physical location in the laser.

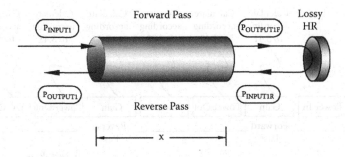

FIGURE 3.17   Single-pass enhancement of the model.

and following that logic, the gain for the second pass in the forward direction would be:

$$g_{FWD2} = \frac{g_0}{1 + \dfrac{(P_{OUTPUT1} + P_{INPUT2})}{P_{SAT}}} \tag{3.49}$$

where P_{INPUT2} is the power exiting the amplifier in the reverse direction reflected from the OC which may, depending on the reflectivity of the OC, be considerably smaller than the output from the previous pass in the reverse direction. This concept (i.e. the use of power in both directions to compute the saturated gain at some point in the laser) is especially important when, for higher accuracy, the amplifier is considered to be several stacked pieces or slices. Ultimately, error will still exist since the gain computed is not constant as photon flux grows during the pass; however, there is a compensating effect in that as the forward power grows, power in the reverse direction decreases and at equilibrium conditions, the product of the forward and reverse power is a constant. This will be considered in the next chapter in which the calculus-based Rigrod approach is examined.

To implement a single-pass model (i.e. where the gain is computed on every pass through the amplifier, and so twice during a round trip through the laser), a spreadsheet is set up in a similar manner to the previous round-trip spreadsheet where each row represents a round trip through the amplifier; however, more columns are required now since the gain, as well as the power, is computed more often, as outlined in Figure 3.18.

As per the previous discussion, gain for the forward and reverse passes is calculated based on both the forward and reverse intensities at that point in the laser amplifier. Gain in the reverse path, for example, is computed based on the output power of the forward pass as well as the input power of the reverse path (the intracavity beam having lost power upon reflection from the HR). As well, whereas in the round-trip approach an amplifier length of $2x$ was used, a length of only x will be used here for computations of power.

Despite the calculation of gain on every pass through the amplifier, gain may still change between passes by an amount large enough such that the gain computed at the start of the next pass is not truly representative of the real gain produced by the

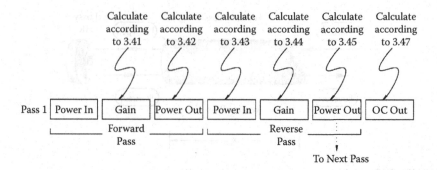

FIGURE 3.18   Spreadsheet formulae for the single-pass model.

amplifier for that pass. In this case, the model may be further enhanced by splitting the amplifier length into two or more pieces and computing the power (and then the saturated gain) at each intermediate junction between each piece. As illustrated in Figure 3.19, each piece is adjacent to the next, so no loss is incurred between the pieces (although a loss still exists at both the HR and the OC during the round-trip through the amplifier).

During the analysis of the enhanced model, it should become apparent that a change is in order in terms of the intensities used to compute the saturated gain according to Equation (3.44). In the simple, round-trip model a value of twice the power which enters the amplifier for that pass was used. With the single-pass enhancement, power in both directions was used to compute the gain for the next pass through the amplifier. Those powers represent the power at the same physical point in the laser which, in the example shown, were just in front of the cavity mirror. Where the amplifier is split into multiple pieces, the intensities used should be the forward and reverse intensities at the same physical point in the amplifier. When computing the gain for the "halfway" point in the reverse pass, for example, one should use the power of the reverse pass (after passing halfway through the amplifier length) plus the forward power at the same point (i.e. the halfway point).

Logically, then, the saturated gain is computed using the values of intra-cavity power as seen in Figure 3.20. So, when computing the gain of the third element in the chain (i.e. the first half of the reverse pass), one would use $P_{OUTPUT1F}$ and $P_{INPUT1R}$ since these represent the reverse and forward powers respectively at that position. When computing the gain of the fourth element, one would use P_{HALF1F} and P_{HALF1R}. Finally, when computing the gain of the first half for the second pass, P_{INPUT1} and $P_{OUTPUT1}$ would be used.

When more than two pieces are used, the same logical arrangement is used to determine which power terms to use when computing gain. Practically, the spreadsheet becomes wider with more columns to represent intermediate gain and power computations.

Increasing the number of logical elements in the model causes the solution to converge to a calculus-based model (where $\Delta x \to dx$).

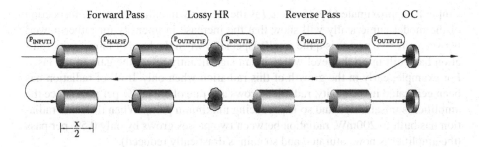

FIGURE 3.19 Half-pass model.

This raises the question "how many pieces are enough?" Logically, as the amplifier is considered as an increasing number of "slices," the thickness of those slices (Δx) decreases until finally becoming an infinitely thin slice, at which point this model converges into a calculus-based model (where $\Delta x \to dx$). To be practical, one can consider the final results of the simulation and compare to the numerical results using previous formulae. For example, when the simulation begins (with an extremely small power), the amplifier delivers a gain of g_0, the small-signal gain which eventually saturates to the threshold value g_{th}. By extension, the output power will also be known since, as the model runs, the intra-cavity power will reach an equilibrium value which cases this saturation. A simple metric for model performance, then, is comparison of the equilibrium values of the model to the established numerical values (namely threshold gain, as computed in Chapter 2) and output power (as computed in Section 3.6).

As an example of "acceptable performance" consider the previous example of a low-gain HeNe laser (Example 3.3) in which intra-cavity power grows so slowly that the total intra-cavity beam power at any time during the growth of this radiation can

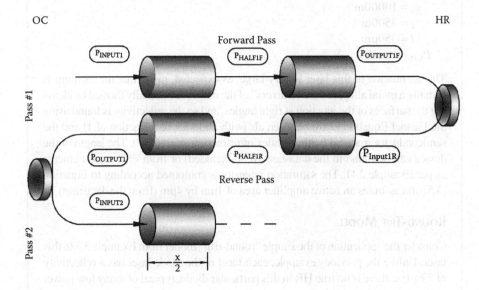

FIGURE 3.20 Logical arrangement of the half-pass model. (Only one pass is shown.)

simply be approximated by $2I$ where I is the intensity in either direction. Inspection of the model numerically will show that the increase in intensity (or rather power) between any two passes is quite small, especially when the level of intra-cavity radiation has built up to the level where it can significantly affect the gain of the laser. For example, early in the growth of this radiation when only 1mW of radiation has been generated in the cavity, radiation grows at a rate of about 4% per pass (since the amplifier is unsaturated and so is producing maximum gain). When the same radiation has built to 200mW, radiation between two passes grows by only 0.5% per pass (the amplifier is now saturated and so gain is drastically reduced).

When applied to a high-gain laser (such as a semiconductor diode type), the basic problem with previous models is that power grows exponentially and the power of the intra-cavity beam between two subsequent passes increases at a high rate (it can be seen to easily double in power in a single pass)—large enough that it cannot be approximated by simply multiplying the intra-cavity radiation by two. Again, comparison of the final equilibrium values for threshold gain and intra-cavity power to numerically-computed values will demonstrate the acceptability of the model so that if further refinement is required, the amplifier can simply be "dissected" into smaller pieces.

EXAMPLE 3.4   COMPARING MODELS FOR A SEMICONDUCTOR LASER

In the same manner as the previous example, the development of power in a semiconductor laser may be modeled; however, the high gain of the medium dictates that an alternative approach be used, as illustrated here.

To begin the model, the following parameters of the diode laser are assumed:

$R_{OC} = 33\%$
$g_0 = 10000 \text{m}^{-1}$
$\gamma = 3500 \text{m}^{-1}$
$l = 350 \mu\text{m}$
$P_{SAT} = 10\text{mW}$

The attenuation of this laser is very large, as expected, given that the medium is actually a metal alloy, and the "mirrors" of the device are actually formed by cleaving the surfaces of the junction at right angles, and so the reflectivity is found using the Fresnel Equation (2.18) between air (with an index of refraction of 1) and the semiconductor material (with an index of refraction around 3.7). The length of the device can be found in the datasheet (if so included) or from FSR measurements as per Example 2.11. The saturation intensity is computed according to Equation (3.5) and assumes an active amplifier area of 1μm by 4μm (from the datasheet).

ROUND-TRIP MODEL

Consider the application of the simple "round-trip" model from Example 3.3 to this laser. Unlike the previous example, each facet of the diode laser has a reflectivity of 33% (i.e. there is no true HR in this particular diode, typical of many low-power

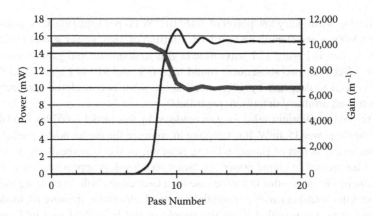

FIGURE 3.21 Model output with a round-trip model.

diode lasers) and so the round-trip model which assumes a lossless HR cannot be used here directly; instead, the OC loss will be assumed to be 33% squared or 10.9% overall reflectivity, but only the portion through one mirror (33%) is used as the output beam. The output from this model is seen in Figure 3.21.

The predicted output power (over 15mW, as evident from the thinner line which indicates power) is quite large compared to the observed results, and a significant amount of "ringing" is predicted in the output (where the output oscillates before settling at an equilibrium value), which is not seen in the lab. The issue here is the assumption that gain remains more or less constant over the entire round trip through the amplifier. What the model does not account for is that the intensity of the beam grows quite rapidly, enough so that the gain saturates to a considerably lower value than expected even during a single pass through the amplifier. To illustrate this, consider the gain and power as computed for several passes through the amplifier in Figure 3.22 (the illustration begins on pass #6).

Round trip (Two passes through amplifier)					Pout OC
Pass#	Round Trip				
	pin	Gain	Actual Gain	Pout (amp)	
	(mW)	(m⁻¹)	(m⁻¹)	(mW)	(mW)
6	0.000294661	1.00E+04	6499.41071	2.79E-02	0.018675
7	0.003035366	9.99E+03	6493.93295	2.86E-01	0.191638
8	0.031148308	9.94E+03	6438.08907	2.82E+00	1.891158
9	0.307383771	9.42E+03	5920.83749	1.94E+01	12.99352
10	2.111931117	7.03E+03	3530.4393	2.50E+01	16.75058
11	2.722593745	6.47E+03	2974.50865	2.18E+01	14.63277
12	2.378371035	6.78E+03	3276.56352	2.36E+01	15.79246

FIGURE 3.22 Spreadsheet for the round-trip model.

Starting with a very low power of $2.46*10^{-10}$W (as per Equation 3.38, assuming an 808nm wavelength and an ULL lifetime of 1ns), power grows rapidly until, as seen in Figure 3.21, saturation begins to dominate and gain decreases rapidly. Between two subsequent round trips (#9, and #10 in Figure 3.22) the gain is seen to decrease by 40%—well above what is considered reasonable (which is an arbitrary definition, regardless).

The final equilibrium value for gain rendered by this model is 6670m^{-1} and the final output power 15.4mW. It is tempting to compare the output power predicted by this model to that of Equation (3.15); however, as you may recall, the simple model for predicting output power was based on a low-loss approximation with a medium of low gain—this is not the case for a laser diode. Still, to compare models, we will continue to analyze potential flaws and continue to improve the model.

Two flaws are revealed when the simple model is applied to a high-gain laser. The first is revealed by the enormous decrease in gain predicted by this model; on one line it is seen to decrease from 5920m^{-1} to 3530m^{-1} and on this same line the input power on this particular pass through the amplifier (the second column in pass #10 in Figure 3.22) is 2.11mW. At the end of this round trip through the amplifier, a power of 16.75mW is achieved—clearly the gain delivered by the amplifier changes during the round trip, and so our assumption of a more or less "constant" gain during a round trip through the amplifier is not logical in the case of a high gain amplifier such as this.

The second flaw (based on the first) is computation of the saturated gain based on Equation (3.7) in which the total intensity that contributes to saturation of gain was assumed to be twice that of the starting intensity for the pass.

Improvements in the model may be made by breaking the amplifier logically into smaller lengths—at first two pieces, where the gain is computed after each half-pass—until the gain is observed to change by a "tolerable" amount. ("Tolerable" is often defined by the performance of the model. When enough segments are considered in the model the ultimate output changes little when another segment is added. For this particular laser little change in the output is evident when the model is upgraded from using a third of a pass to using a quarter of a pass.)

SINGLE-PASS MODEL

Next, a single-pass model will be used with the same laser where the gain is computed after each single pass (both forward and reverse) through the amplifier, with the results of the model seen in Figure 3.23.

The results of this model are considerably closer to those observed in the lab, predicting an output power of 2.28mW (which is lower than observed but considerably more accurate than the previous model). Examination of several passes in Figure 3.24 still reveals that the jump in gain and power between subsequent passes is considerably smaller than in the previous model. Consider, for example, that gain between the forward and reverse passes for pass #9

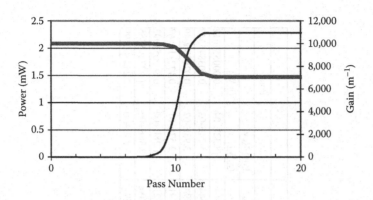

FIGURE 3.23   Model output with a single-pass model.

(shown in third and seventh columns) is 11%—a massive improvement over the previous model (for such a simple enhancement), yet there is still room for improvement by considering the amplifier to be many more pieces.

Again, comparison of the equilibrium values determined using this model with the numerical values computed previously shows much better agreement: the output power achieves an equilibrium value of 2.28mW.

Not evident from Figure 3.24 is the manner in which gain is computed; that is, by summing the power of the forward and reverse passes as the same physical point in the laser rather than the simplification of $2I$ as used in Equation (3.7). The intensity used in this case is the forward power plus the power in the reverse direction as follows:

$$g = \frac{g_0}{1 + \frac{(I_{FWD} + I_{REV})}{I_{SAT}}}$$

The saturated gain for the reverse pass, for example, was computed by using the output power from the forward pass (the column labeled P_{OUT}-FWD) and the input power from the reverse pass (the column labeled P_{IN}-REV) since these powers occur in opposite directions at the same physical location—inside the amplifier immediately adjacent to the rear reflector. Of course the input to the reverse pass is just the output from the first pass multiplied by the reflectivity of the reflector (0.33, in this case). The same is true for the gain on the forward pass which is computed from the output power of the reverse pass (the column labeled P_{OUT}-REV) and the input power for the forward pass (the column labeled P_{IN}-FWD). The output power from the reverse pass is taken from the previous pass and so the numerical row reference is the current row minus one.

Output power is taken arbitrarily from the power through one optic (in this case, the rear optic) and the gain shown on the graph is that of the forward pass only.

Single-pass through amplifier									
Pass#	Forward Pass				Reverse Pass				Pout
	Pin-Fwd (mW)	Gain (m⁻¹)	Actual Gain (m⁻¹)	Pout-Fwd (mW)	Pin-Rev (mW)	Gain (m⁻¹)	Actual Gain (m⁻¹)	Pout-Rev (mW)	(mW)
6	2.6655E-05	1.00E+04	6499.852791	2.59E-04	8.55637E-05	1.00E+04	6499.655164	8.32E-04	0.000557612
7	0.000184012	1.00E+04	6498.983835	1.79E-03	0.000590507	1.00E+04	6497.620644	5.74E-03	0.003845547
8	0.001269031	9.99E+03	6492.996255	1.23E-02	0.004063885	9.98E+03	6483.648094	3.93E-02	0.026336049
9	0.008690896	9.95E+03	6452.230852	8.31E-02	0.027437049	9.89E+03	6390.929791	2.57E-01	0.172110567
10	0.056796487	9.70E+03	6195.862202	4.97E-01	0.16391756	9.38E+03	5880.302121	1.28E+00	0.860051092
11	0.28381686	8.64E+03	5144.928838	1.72E+00	0.567018336	8.14E+03	4639.838706	2.88E+00	1.927265855
12	0.635997732	7.40E+03	3900.547411	2.49E+00	0.821986496	7.51E+03	4011.53697	3.35E+00	2.242360399

FIGURE 3.24 Single-pass model.

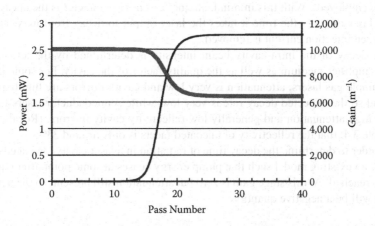

FIGURE 3.25   Third-pass model.

HALF- AND THIRD-PASS MODELS

The model is improved further by considering the amplifier length to be two, three, or more concurrent pieces with the output of the third-pass model (i.e. where gain is computed on a single forward pass through the amplifier three times) in Figure 3.25. The accuracy of the model is markedly improved here and the predicted output power of 2.78mW matches reasonably well with the actual observed output power (as well as the analytical solution that follows in the next chapter when one considers that there are actually two output beams of equal output power).

Examining a few lines of the spreadsheet (not shown here due to the sheer size, since it is over twenty columns in width), one would notice that gain changes a maximum of about 5% between any two successive amplifier elements—a reasonable approximation for this laser. (Care must be taken, however, to compute gain properly from power figures at the same physical point in the cavity.) The actual number of elements into which an amplifier must be dissected is based on this change—the third-pass model was deemed sufficient since the output power predicted by the half-pass model was 2.75mW and by this third-pass model 2.78mW. Breaking the amplifier into more pieces will result in a negligible change in the predicted output power.

3.11   DETERMINING CAVITY DECAY PARAMETERS

In the previous section, the model was applied to a semiconductor diode laser. Of particular interest is the ability of the model to predict the rise-time of the diode—the time for the output of the diode to reach 90% of its maximum power. For a diode laser used for communications this is one of the most important parameters since it dictates the maximum modulation rate (and hence the maximum data rate which may be employed). With this in mind, another essential parameter is the decay time of the laser output—the time it takes the laser output to decay from maximum to 10% when injection current is removed.

The decay of the intra-cavity beam intensity is determined by the attenuation of the amplifier medium as well as the quality factor of the cavity. For some lasers, for example gas lasers, attenuation is very low and cavity mirrors are highly reflective and so the expected decay rate is very low, while semiconductor lasers feature a very high attenuation and generally low reflectivity cavity mirrors. (Recall, from Example 3.4, that the reflectivity of uncoated facets is only around 33%.)

In order to determine the decay time of radiation in a laser cavity, one needs only modify an existing model such that pump energy ceases at some point after equilibrium is reached. At this stage gain is zero but attenuation still present, so the net gain shown will be a negative quantity.

> The rate of decay of radiation inside a laser cavity is the basis for cavity ring-down spectroscopy (CDRS).

EXAMPLE 3.5   DECAY IN A HENE LASER

Following Example 3.3 in which the power buildup in a HeNe laser was examined—the simplest model is used here for illustration—decay in a high-Q cavity can be illustrated by setting the gain to zero at some point after equilibrium is reached (referring back to the original example, any time after about 1200 passes). This is accomplished by simply changing the spreadsheet so that the saturated gain is zero at some point (in this case, at pass #1500, well after the laser has achieved an equilibrium state), as seen in Figure 3.26. The resulting spreadsheet shows the intra-cavity power decay from a constant level of 337mW due to loss at both the OC (1% is lost per reflection) as well as attenuation

	A	B	C	D	E	F
10	Pass#	Pin	Gain	Actual Gain	Pout(amp)	Pout
1508	1496	333.4143007	2.78E-02	0.02284466	3.37E+02	3.367826
1509	1497	333.4147383	2.78E-02	0.02284463	3.37E+02	3.36783
1510	1498	333.4151715	2.78E-02	0.0228446	3.37E+02	3.367834
1511	1499	333.4156004	2.78E-02	0.02284457	3.37E+02	3.367839
1512	1500	333.416025	0.00E+00	−0.005	3.33E+02	3.326833
1513	1501	329.3564828	0.00E+00	−0.005	3.29E+02	3.286327
1514	1502	325.3463681	0.00E+00	−0.005	3.25E+02	3.246314
1515	1503	321.3850789	0.00E+00	−0.005	3.21E+02	3.206788
1516	1504	317.4720208	0.00E+00	−0.005	3.17E+02	3.167744

FIGURE 3.26   Decay in a cavity.

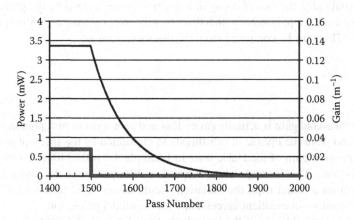

FIGURE 3.27 Decay of cavity radiation in a HeNe laser.

(which is seen here as a negative gain). The reason for separating "gain" and "actual gain" in the spreadsheet of Example 3.3 should now be apparent.

The output from the model is seen graphically in Figure 3.27, in which the equilibrium values are seen at pass 1400 and pumping (and hence gain) stops abruptly at pass 1500. Owing to the high quality of the cavity, intensity is seen to fall to $1/e$ (36.8% of the equilibrium value) in about eighty passes.

As a second example, the same gain switching is applied to the semiconductor laser of Example 3.4 in which the third-pass model is used for accuracy. In this case, gain is set to zero at pass #30 (Figure 3.28).

Decay in a semiconductor laser is very rapid and over 90% of cavity radiation is lost in one pass—not surprising given the fact that two-thirds is lost on each reflection from a cavity facet and the enormous attenuation of the medium causes what remains to decrease rapidly. (A reduction to about one-third of the initial value occurs in about three-quarters of a single pass.)

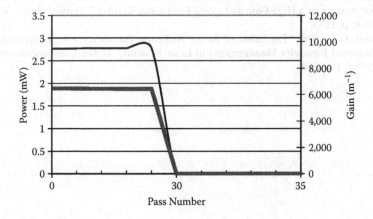

FIGURE 3.28 Decay of cavity radiation in a semiconductor laser.

Mathematically, the rate of decay of a cavity is characterized by the photon life-time of the cavity, a parameter which describes the average time a photon spends in the cavity. This may be computed from the threshold gain as:

$$\tau_{CAVITY} = \frac{1}{c \times g_{th}} \qquad (3.50)$$

Since the threshold gain is actually cavity loss and is in units of m^{-1}, photon lifetime can be found from the inverse of this threshold multiplied by the speed of light.

The threshold gain of the HeNe laser in Example 3.2 is $0.0340m^{-1}$ so the cavity lifetime is 98ns. This corresponds to a travel of 29m and since the cavity is 22cm long (44cm on a round trip), the radiation should decay in 67 passes after gain is switched to zero—in excellent agreement with the model prediction.

Now, the threshold gain of the laser diode from Example 3.4 is $6667m^{-1}$ and the corresponding cavity lifetime is $5.0*10^{-13}$sec. With an n value of 3.7 it is expected that radiation would travel only 41μm in the medium; however, this assumes a completely distributed system, whereas the physical reality of the system is that a large loss occurs at each mirror. Regardless, decay occurs in less than one pass through the device.

REFERENCES

1. NIST: National Institute of Standards and Technology, Physical Measurements Laboratory, 2013. *Atomic Spectral Database*, http://physics.nist.gov/PhysRefData/ASD/lines_form.html
2. Schulz-DuBois, E.O. 1964. "Pulse Sharpening and Gain Saturation in Traveling-Wave Masers." *Bell System Technical Journal*, March 1964, p. 625.
3. Rigrod, W.W. 1965. "Saturation Effects in High-Gain Lasers." *Journal of Applied Physics*, Vol. 36, No. 8, p. 2487.
4. Milonni, P.W. 1981. "Saturation of Anomalous Dispersion in cw HF Lasers." *Applied Optics*, Vol. 20, No. 9, p. 1571.
5. Carroll, D.L. 1994. "Effects of a Nonhomogeneous Gain Saturation Law on Predicted Performance of a High-Gain and a Low-Gain Laser Systems." *Applied Optics*, Vol. 33, No. 9, p. 1673.
6. Ernst, G.J. 1978. "The Effect of Radial Radiation Transport on the Interpretation of Saturation Parameter Measurements in Laser Systems." *Optics Communications*, Vol. 25, No. 3, p. 368.

4 Analytical Solutions

This chapter presents primarily the calculus-based approach to laser models that yield analytical results. Included are the Rigrod approach and the development of a formula for the determination of optimal OC transmission. In this chapter, the methodology of development of these approaches is more important than the results themselves since these may be applied to any specific laser configuration desired. For this reason, the Rigrod approach is demonstrated with two specific configurations. These formulae will be modified later when losses such as re-absorption loss are considered in Chapter 5.

4.1 THE RIGROD APPROACH

The "gold standard" model for the prediction of output power and the optimization of laser optics is the Rigrod approach. Presented here is a somewhat simplified approach based on the same logic as the original paper by Rigrod.[1]

The approach begins, as per that employed in the previous chapter (in Section 3.4), with the fact that there are two beams traversing the laser cavity in opposite directions, denoted I_- and I_+. The forward (I_+) path begins with an intensity of I_1 and ends with an intensity of I_2. A portion of that beam is then transmitted by mirror R_2 (the OC) to become the output beam and the remaining quantity, reduced in intensity due to the loss at that mirror, becomes the beginning of the reverse (I_-) path denoted I_3. That beam is amplified to become I_4, loses a portion of its intensity at R_1, and the cycle is repeated. The beam intensity as a function of position is diagrammed in Figure 4.1.

The power increase of each beam (dI/dz) is a function of gain and beam intensity according to:

$$\frac{dI_+}{dz} = g(z)I_+ \tag{4.1}$$

and

$$\frac{dI_-}{dz} = g(z)I_- \tag{4.2}$$

where $g(z)$ is the gain as a function of position in the amplifier (which, of course, is the saturated gain, which varies due to changes in the intensity of the intra-cavity beams traversing the laser). As well, the product of these two beams traversing the amplifier is assumed to be constant at any point in the gain medium, and so

$$I_+I_- = C \tag{4.3}$$

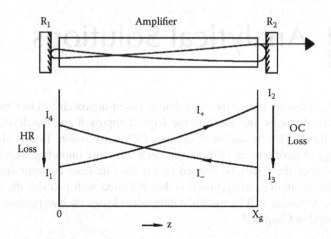

FIGURE 4.1 Circulating intensities in a laser.

Considering only the beam in the forward direction, substituting the saturated gain formula of Equation (3.8) yields:

$$\frac{dI_+}{dz} = \frac{g_0}{1 + \dfrac{I_+ + I_-}{I_{SAT}}} I_+ \tag{4.4}$$

Eliminating I_- with Equation (4.3),

$$\frac{dI_+}{dz} = \frac{g_0}{1 + \dfrac{I_+ + \dfrac{C}{I_+}}{I_{SAT}}} I_+ \tag{4.5}$$

The equation is now rearranged:

$$dI_+ \left(1 + \frac{I_+}{I_{SAT}} + \frac{\dfrac{C}{I_+}}{I_{SAT}}\right)\frac{1}{I_+} = g_0 dz \tag{4.6}$$

The terms are now integrated with the limits of the forward beam being I_1 and I_2, and the limits of the gain length terms being 0 and x_g (i.e. the length of the gain element):

$$\int_{I_1}^{I_2}\left(\frac{1}{I_+} + \frac{1}{I_{SAT}} + \frac{C}{I_+^2 I_{SAT}}\right)dI_+ = \int_0^{x_g} g_0 dz \tag{4.7}$$

the results being

$$\ln(I_2) - \ln(I_1) + \frac{I_2}{I_{SAT}} - \frac{I_1}{I_{SAT}} + \frac{C}{I_{SAT}}\left(\frac{-1}{I_2} - \frac{-1}{I_1}\right) = g_0 x_g \quad (4.8)$$

or, rearranged

$$\ln\left(\frac{I_2}{I_1}\right) + \frac{1}{I_{SAT}}(I_2 - I_1) - \frac{C}{I_{SAT}}\left(\frac{1}{I_2} - \frac{1}{I_1}\right) = g_0 x_g \quad (4.9)$$

In order to solve the equation for I_2 (the ultimate goal), we must first substitute for I_1. Using Equation (4.3), and knowing that

$$I_3 = \frac{C}{I_2} = R_2 I_2 \quad (4.10)$$

the constant is then found to be

$$C = R_2 I_2^2 \quad (4.11)$$

Similarly, we can relate I_1 and I_4 as follows:

$$I_1 = \frac{C}{I_4} = R_1 I_4 \quad (4.12)$$

$$C = R_1 I_4^2 \quad (4.13)$$

From Equations (4.11) and (4.13),

$$I_4 = \sqrt{\frac{R_2}{R_1}} I_2 \quad (4.14)$$

From Equations (4.12) and (4.14),

$$I_1 = \frac{C}{I_4} = \frac{R_2 I_2^2}{\sqrt{\frac{R_2}{R_1}} I_2} = \sqrt{R_1 R_2} I_2 \quad (4.15)$$

Finally, substituting for I_1 in Equation (4.9) yields

$$\ln\left(\frac{1}{\sqrt{R_1 R_2}}\right) + \frac{I_2}{I_{SAT}}(1 - \sqrt{R_1 R_2}) - \frac{I_2}{I_{SAT}}\left(R_2 - \sqrt{\frac{R_2}{R_1}}\right) = g_0 x_g \quad (4.16)$$

And rearranging to solve for I_2

$$I_2 = \frac{I_{SAT}(g_0 x_g + \ln(\sqrt{R_1 R_2}))}{1 + \sqrt{\frac{R_2}{R_1}} - R_2 - \sqrt{R_1 R_2}} \tag{4.17}$$

This is the intra-cavity intensity. To compute the intensity exiting the output coupler this is multiplied by the fraction transmitted by the OC:

$$I_{OC} = (1 - R_2)I_2 \tag{4.18}$$

One key feature of the Rigrod formulation is the assumption that attenuation of the amplifier medium (i.e. γ) is zero. To include an attenuation term, one often expresses attenuation as a dimensionless quantity and hence it becomes a "fixed" loss easily represented as R_1 (which, in a practical standing-wave laser, is usually 1.00 or close to it). This is a valid approach since γ is usually a function of amplifier length alone (and thus is not a function of, say, intensity in the cavity—this is especially true in four-level lasers). In this manner, we would express the attenuation loss as follows:

$$R_1 = e^{-2\gamma x_a} \tag{4.19}$$

which follows the logic introduced in Section 1.7 and continued in Example 2.3. For small values of attenuation, this relation may be approximated as:

$$R_1 = 1 - 2\gamma x_a \tag{4.20}$$

In the Rigrod solution of Equation (4.17), then, the value R_1 now represents the portion kept in the cavity after the loss due to attenuation and, perhaps, reflection loss at R_1 were it not 100% reflective. Where the HR has a significant loss, R_1 can incorporate both that loss and attenuation simply by multiplying the above quantity by the actual reflectivity of that mirror. In other words,

$$R_1 = R_{HR}e^{-2\gamma x_a} \tag{4.21}$$

Or, in the simplified form,

$$R_1 = R_{HR}(1 - 2\gamma x_a) \tag{4.22}$$

where R_{HR} is the actual reflectivity of the HR and R_1 is now the "apparent" reflection, which incorporates both actual reflectivity and attenuation in the amplifier.

> The basic Rigrod formula assumes $\gamma = 0$ so attenuation loss must be mathematically "lumped" onto the mirrors.

EXAMPLE 4.1 PREDICTING OUTPUT POWER USING THE RIGROD APPROACH

In this example, we compute the expected output power for the same HeNe laser outlined in Example 2.1. To recap, the parameters of the laser are as follows:

$R_{HR} = 100\%$
$R_{OC} = 99.0\%$
$x_a = 22\text{cm}$
$x_g = 18\text{cm}$
$\gamma = 0.005\text{m}^{-1}$

Since the Rigrod formula of Equation (4.17) does not include the absorption (γ) of the amplifier medium, this may be represented in the same manner as a "fixed" loss, such as a mirror, using Equation (4.20) as

$$R_1 = 1 - 2(0.005m)(0.22m)$$

or $R_1 = 0.9978$ to represent this loss. The Rigrod formula of Equation (4.17) is then combined with Equation (4.18) to determine the actual output power. Using power instead of intensity, and using the previously calculated value for saturation power from Example 3.1—with a bore of 0.8mm the saturation power was calculated to be 152mW—the predicted output power is:

$$P_{OC} = \frac{(1 - 0.99)(152mW)(0.15m^{-1} \times 0.18m + \ln(\sqrt{0.99 \times 0.9978}))}{1 + \sqrt{\dfrac{0.99}{0.9978}} - 0.99 - \sqrt{0.99 \times 0.9978}} = 2.60mW$$

As expected, this compares quite favorably to the simple algebraic model based on the gain saturation expression of Example 3.2 and the single-pass spreadsheet-based model of Example 3.3. Where a high gain device is considered (with high coupling losses), the Rigrod model offers a substantial advantage over other models.

Where a laser has an extraordinarily high attenuation, a different approach is required. In the previous example, attenuation was "lumped" onto one mirror (the HR), but where cavity optics have a high loss and attenuation is large this approach will not yield an accurate answer. The approach taken here is to "spread out" attenuation loss between both cavity optics, in effect, computing a "total effective" reflectivity of the cavity optics. Assuming, as we did in Section 3.10, a cavity consisting of two identical mirrors, the total reflectivity of the mirrors can be represented as:

$$R_{TOTAL} = R_1 R_2 e^{-2\gamma x_a} \quad (4.23)$$

And so each mirror has an effective reflectivity of:

$$R_1 = R_2 = \sqrt{R_{TOTAL}} \tag{4.24}$$

This reflectivity is then used with the basic Rigrod formula of Equation (4.17) to determine the intra-cavity power. Since the cavity mirrors are considered to be equal, the Rigrod formula now simplifies as follows:

$$I_2 = \frac{I_{SAT}(g_0 x_g + \ln(R_1))}{1 + 1 - R_1 - R_1} \tag{4.25}$$

The actual output power from the OC is then computed using Equation (4.18) and employing the actual reflectivity of the optic since we are concerned only about the portion transmitted through the optic.

As an aside, this same technique can be applied to any laser including that from the previous example, which has a relatively low attenuation (i.e. it is a "general" technique). In this case, a difference of only a few percent will be seen in the solutions employing either technique.

EXAMPLE 4.2   APPLICATION TO A HIGH-GAIN LASER

In this example, we compute the expected output power for the same semiconductor laser outlined in the numeric Example 3.4. To recap, the parameters of the laser are as follows:

$R_{OC} = 33\%$

$g_0 = 10000\text{m}^{-1}$

$\gamma = 3500\text{m}^{-1}$

$l = 350\mu\text{m}$

$P_{SAT} = 10\text{mW}$

To reinforce the necessity of the "total effective reflectivity" approach, the same approach is employed here as employed in Example 4.1, in which attenuation (γ) of the amplifier medium is incorporated into the HR loss using Equation (4.19). That form is specifically used since the large attenuation does not allow use of the simplified form of Equation (4.21). For the purposes of the Rigrod analysis, R_1 is then:

$$R_1 = 0.33e^{(-2 \times 3500\text{m}^{-1} \times 350 \times 10^{-6}\text{m})} = 0.0285$$

The Rigrod formula of Equation (4.17), combined with Equation (4.18), is then used in a straightforward manner to predict the output power as:

$$P_{OC} = \frac{(1 - 0.33)(10mW)((10000m^{-1} \times 350 \times 10^{-6}m) + \ln(\sqrt{0.33 \times 0.0285}))}{1 + \sqrt{\frac{0.0285}{0.3}} - 0.33 - \sqrt{0.33 \times 0.0285}} = 9.02mW$$

The solution presented is simplified and optimized for a homogeneously broadened medium. It is possible to formulate Rigrod solutions for purely non-homogeneous media as well as optimize the solution for various physical situations, but analytical solutions become difficult to obtain—this was the original reason for use of a homogeneous approximation in this solution. Still, one may use approximations to simplify various formulations[2] allowing analytical solutions.

4.2 RING LASERS

As a second example of the application of Rigrod theory to a particular system (which illustrates the methodology involved), consider the ring laser shown in Figure 4.2, in which lasing occurs in one direction only around the ring by the inclusion of an optical diode into the ring. In this laser mirrors M1 through M3 are high reflectors and the optical diode (which consists of several optical elements collectively forming the device) has a transmission of T_{OD}.

The analysis begins in much the same way as the previous analysis for the standing wave laser but is simplified since only a single beam (of intensity I) exists,

---

This clearly does not agree well with previous models, nor does it agree with observed power output of the device (which is actually rated for an output of 5mW). The issue is the approach of "lumping" all attenuation loss onto one mirror—the extraordinarily high attenuation value requires, logically, that losses be distributed between both cavity mirrors. To accomplish this, a figure is computed to represent the total effective cavity reflectivity as follows:

$$R_{TOTAL} = R_1 R_2 e^{-2\gamma\alpha} = 0.33^2 e^{(-2\times3500m^{-1}\times350\times10^{-6}m)} = 9.397\times10^{-3}$$

And so each mirror has an effective reflectivity of

$$R_1 = R_2 = \sqrt{9.397\times10^{-3}} = 0.0969$$

The intra-cavity power is then computed according to Equation (4.25) as:

$$P_2 = \frac{(10mW)((10000m^{-1}\times350\times10^{-6}m)+\ln(0.0969))}{2-2(0.0969)} = 6.46mW$$

Finally, the output power is computed according to Equation (4.18) using the true reflectivity of the OC (not the "effective" value since we are concerned only with the portion that is transmitted through the OC while the "effective" value incorporates losses at both cavity mirrors as well as attenuation). Using an OC transmission of 0.67, this gives a predicted output power of 4.32mW, which predicts a higher output than the previous numerical models and approaches the rated power of the device (which is rated at 5mW).

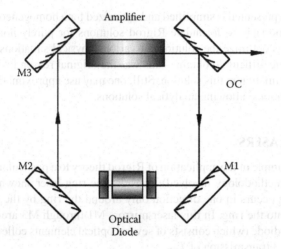

FIGURE 4.2 A ring laser.

rotating around the ring in a clockwise direction. The power increase of the beam (dI/dz) is a function of gain and beam intensity according to:

$$\frac{dI}{dz} = g(z)I \tag{4.26}$$

Using the saturated gain formula of Equation (3.6), which applies to a unidirectional beam, yields:

$$\frac{dI}{dz} = \frac{g_0}{1 + \dfrac{I}{I_{SAT}}} I \tag{4.27}$$

The equation is now rearranged:

$$dI\left(1 + \frac{I}{I_{SAT}}\right)\frac{1}{I} = g_0 dz \tag{4.28}$$

The terms are now integrated with the limits of the forward beam being I_1 and I_2 (these intensities defined as the input to the amplifier and the output from the

An optical diode uses a magneto-optic effect (using a Faraday rotator) to allow light to travel in one direction only.

amplifier, respectively), and the limits of the gain terms being 0 and x_g (the length of the gain element):

$$\int_{I_1}^{I_2}\left(\frac{1}{I}+\frac{1}{I_{SAT}}\right)dI = \int_0^{x_g} g_0 dz \tag{4.29}$$

The results of the integration are

$$\ln(I_2) - \ln(I_1) + \frac{I_2}{I_{SAT}} - \frac{I_1}{I_{SAT}} = g_0 x_g \tag{4.30}$$

or, rearranged

$$\ln\left(\frac{I_2}{I_1}\right) + \frac{1}{I_{SAT}}(I_2 - I_1) = g_0 x_g \tag{4.31}$$

A simple substitution is made for intensity I_1 as follows:

$$I_1 = R_x R_{OC} I_2 \tag{4.32}$$

where R_{OC} is the reflectivity of the OC and R_x is the reflectivity of the other three optics (and the transmission of any other intra-cavity optics such as the optical diode) combined; in other words,

$$R_x = R_1 R_2 R_3 T_{OD} \tag{4.33}$$

So, Equation (4.31) simplifies as:

$$\ln\left(\frac{I_2}{I_2 R_x R_{OC}}\right) + \frac{1}{I_{SAT}}(I_2 - I_2 R_x R_{OC}) = g_0 x_g \tag{4.34}$$

$$\frac{I_2}{I_{SAT}}(1 - R_x R_{OC}) = g_0 x_g + \ln(R_x R_{OC}) \tag{4.35}$$

Solving for intensity I_2,

$$I_2 = I_{SAT}\left(\frac{1}{1 - R_x R_{OC}}\right)(g_0 x_g + \ln(R_x R_{OC})) \tag{4.36}$$

Finally, solving for output power from the OC,

$$I_{OC} = (1 - R_{OC})I_{SAT}\left(\frac{1}{1 - R_x R_{OC}}\right)(g_0 x_g + \ln(R_x R_{OC})) \tag{4.37}$$

EXAMPLE 4.3   APPLICATION TO A RING LASER

Using the laser in Figure 4.2 as an example, consider this laser as having the following parameters:

R_1 to R_3 = 99.95% (all are high reflectors at an angle of 45 degrees)
R_{OC} = 99.0%
T_{OD} = 97.0% (a low-loss model was chosen)
X_g = 35cm
γ = 0.005m^{-1}

The first question to be answered is, "Will it oscillate?" The gain threshold equation is formulated in the manner of Equation (2.11), but since there is only a unidirectional beam path, gain occurs only once per round trip:

$$g_{th} = \gamma + \frac{1}{x}\ln\left(\frac{1}{R_1 R_2 R_3 R_{OC} T_{OD}}\right)$$

Substitution of the known parameters for the laser yields an answer of 0.125m^{-1}. Knowing that the typical gain of a HeNe amplifier is 0.15m^{-1}, we conclude that the ring laser will indeed oscillate.

We now use Equation (4.37), but first the value of R_x (the "combined" reflectivity of the optical elements in the cavity) according to Equation (4.33) is:

$$R_x = (0.9995^3)(0.97)$$

or 0.9685. Using Equation (4.37), the output power is now computed (with saturation power assumed to be 152mW as in the previous example) to be:

$$P_{OC} = (0.01)152mW\left(\frac{1}{1-0.99\times0.9685}\right)(0.15\times0.35m + ln(0.99\times0.9685))$$

or 0.385mW. Not surprisingly, the output power is low given the loss of the optical diode (which was chosen as it is a low loss model—most optical diodes have a transmission of about 85%).

4.3   OPTIMAL OUTPUT COUPLING

Assuming that the gain medium is of fixed type and length, the logical parameter to optimize is the transmission of the output coupler. An expression for output power was developed in Section 3.5 (Equation 3.31) for small values of OC transmission (<10%) as:

$$P_{OUT} = \frac{1}{2}P_{SAT}T_{OC}\left(\frac{2x_g g_0}{2\gamma x_a + T_{OC}} - 1\right) \qquad (4.38)$$

which is then expanded as:

$$P_{OUT} = \frac{1}{2} P_{SAT} T_{OC} \left(\frac{2x_g g_0}{2\gamma x_a + T_{OC}} \right) - \frac{1}{2} P_{SAT} T_{OC} \qquad (4.39)$$

To find the optimal value of OC transmission, this equation is differentiated (with the help of a calculus identity) with respect to the transmission of the OC and the differential set to zero:

$$\frac{\partial P_{OUT}}{\partial T_{OC}} = \frac{1}{2} P_{SAT} \left(\frac{1}{2\gamma x_a + T_{OC}} \frac{\partial (2x_g g_0)}{\partial T_{OC}} - \frac{2x_g g_0}{(2\gamma x_a + T_{OC})^2} \frac{\partial (2\gamma x_a + T_{OC})}{\partial T_{OC}} - 1 \right) = 0 \quad (4.40)$$

$$\frac{\partial P_{OUT}}{\partial T_{OC}} = \frac{1}{2} P_{SAT} \left(\frac{2x_g g_0}{2\gamma x_a + T_{OC}} - \frac{2x_g g_0 T_{OC}}{(2\gamma x_a + T_{OC})^2} - 1 \right) = 0 \qquad (4.41)$$

which is then solved for the optimal value of OC transmission (using a mathematical solver) to yield:

$$T_{OC} = \sqrt{2x_g g_0 (2\gamma x_a)} - 2\gamma x_a \qquad (4.42)$$

One could have used the Rigrod model of Equation (4.17) from earlier in this chapter as well to derive an expression for optimal OC coupling by differentiating the output power with respect to the reflectivity of the OC (i.e. R_{OC}), although this will certainly yield a more complex mathematical formulation. Similarly, the expression for the optimal OC transmission of the ring laser from Section 4.2 can be developed starting with the Rigrod expression for output power.[3]

EXAMPLE 4.4   PREDICTING OPTIMAL CAVITY OPTICS

Consider the HeNe laser of Example 2.1 with the following parameters:

$R_{HR} = 100\%$
$R_{OC} = ?$
$X_a = 22\text{cm}$
$X_g = 18\text{cm}$
$\gamma = 0.005\text{m}^{-1}$

In theory, one could start with the Rigrod expression for output power and differentiate this with respect to R_{OC} for a solution.

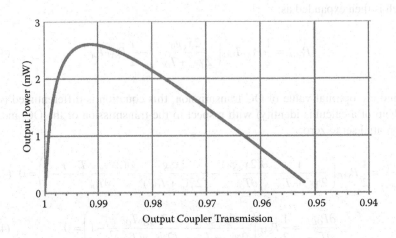

FIGURE 4.3   Output power as a function of OC transmission.

Knowing the small-signal gain of the HeNe amplifier is 0.15m⁻¹, the optimal OC transmission is computed according to Equation (4.42) to be:

$$T_{OC} = \sqrt{(2 \times 0.22m \times 0.15m^{-1})(2 \times 0.005m^{-1} \times 0.18m)} - 2 \times 0.005m^{-1} \times 0.22m$$

The optimal value of OC transmission is hence 0.91% (and so the optimal OC reflectivity is 99.09%, which agrees quite well with results from previous models). Of course, this leads to the question, "How precise does this value need to be?" In Figure 4.3 the output power of this laser is graphed as a function of OC transmission (using Equation 3.27, from which this solution, which began with Equation 4.38, was derived). As expected, the peak output agrees with the numerical solution of 99.1% reflectivity. The graph also shows that even when the transmission of the OC is double that of the optimal value, the output power is still 85% of the maximum value, illustrating the sensitivity of output power to transmission of the OC.

REFERENCES

1. Rigrod, W.W. 1965. "Saturation Effects in High-Gain Lasers." *Journal of Applied Physics*. Vol. 36, No. 8.
2. Casperson, L.W. 1980. "Laser Power Calculations: Sources of Error." *Applied Optics*, Vol. 19, No. 3, p. 422.
3. Eckbreth, A.C. 1975. "Coupling Considerations for Ring Resonators." *IEEE Journal of Quantum Electronics*, Sept 1975, p.796.

5 Thermal Issues

The purpose of this chapter is to outline the various ways in which temperature affects lasers, and to model those effects. Four-level lasers are generally unaffected by temperature change, but four-level lasers with a LLL particularly close to ground level are affected. It is ironic, then, that these quasi-three-level lasers, which offer the highest efficiencies, are also those most affected. This chapter examines these materials in detail, and models are presented to determine the effect of temperature change on laser output.

In addition to affecting certain solid-state lasers, temperature also affects the emission wavelength of semiconductor lasers, for example, and the resulting change is quite notable. The output of many diode lasers will drift 0.25nm for every temperature change of 1°C—a change that has huge ramifications when the diode is pumping a solid-state laser with a very narrow absorption peak.

This chapter is quite relevant, then, to DPSS design, both from the standpoint of the behavior of the amplifier and the diode pump source.

5.1 THERMAL POPULATIONS AND RE-ABSORPTION LOSS

In a three-level laser, such as a ruby, loss due to absorption of photons can be quite large. Imagine an unpumped ruby amplifier such that essentially all atoms of chromium will be at ground state. These atoms will readily absorb laser radiation at the same wavelength as the wavelength of emission (in ruby, 694.3nm), pumping chromium ions to what is normally the ULL in the process. Absorption under these conditions will be quite large. In fact, absorption will be as large as gain, so that if a particular ruby rod exhibits a small-signal gain (g_0) of, say, 25m^{-1} when pumped to 100% inversion, that same rod would exhibit an absorption loss (γ) of 25m^{-1} when unpumped. This is expected in a three-level laser system. Absorption is not constant, though, since as pumping ensues more chromium ions reach the ULL and so fewer are available at the lower level (ground, in this case) to absorb photons at all. When the break-even point is reached (where there are as many ions in the ULL as in the LLL), absorption equals gain (not that this is sufficient to ensure lasing, since inversion must still build to the point where the gain it produces overcomes losses in the cavity). As inversion grows further, absorption decreases until, at 100% inversion, no ions exist in the lower level at all to absorb photons. This does not imply $\gamma = 0$ in the threshold gain equation, though, since losses such as scattering will still occur in the material, but true absorption from the LLL is zero.

If the ruby amplifier were not pumped in the traditional manner, but were instead subject to large values of cavity radiation at the emission wavelength, it would quickly absorb that radiation, but in the process, ions would effectively be "pumped" to the ULL and the material would become transparent. The absorption, then, is *saturable,* meaning the material will absorb cavity radiation until it saturates and

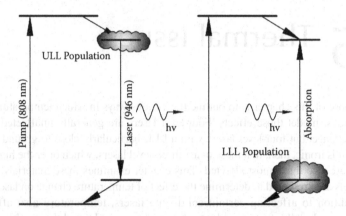

FIGURE 5.1 Emission and re-absorption in a laser.

becomes transparent to that wavelength. (This same concept will be explored in the next chapter in the discussion of passive Q-switches.)

In a three-level laser, it is obvious that absorption at the lasing wavelength is enormous and so has a drastic effect on the minimum pump power required to achieve lasing action. However, four-level lasers (at least "ideal" ones) have no such problems with absorption of laser energy. The presence of a discrete LLL that is essentially empty (i.e. where the LLL has an incredibly short lifetime so that a significant population cannot build up in that level) means that there are no atoms at the lower level available to absorb that wavelength. This is the primary reason for the high efficiency (and low pump thresholds) of a four-level laser system.

Now, while absorption in an unpumped three-level medium is quite large, similar effects are seen in four-level lasers where the LLL is so close to the ground state that that level is thermally populated. The thermal population of the LLL causes the minimum pump power required to bring the laser to threshold to increase, since for every atom in the LLL one must pump an atom into the ULL to overcome this. Such lasers are called *quasi-three-level lasers*.

The overall effect of thermal population at the LLL is most often represented as a loss called *re-absorption loss,* in which intra-cavity radiation is absorbed, or rather re-absorbed, by the LLL population. As depicted in Figure 5.1, a "normal" four-level laser has a large population of atoms (the "cloud" in the left diagram) in the ULL and essentially none in the LLL (owing to the fact that in a practical four-level laser the LLL has a much shorter lifetime than the ULL). This leads to a large amount of emission at the lasing wavelength but essentially no absorption, since atoms must be present in the LLL to be able to absorb radiation. In a laser with a significant LLL population (the "cloud" in the right diagram), a significant population of atoms in the LLL leads to a large absorption of photons at the lasing wavelength, as per Equation (1.4).

For the majority of four-level lasers, thermal population of the LLL is insignificant.

5.2   QUASI-THREE-LEVEL SYSTEMS

Where a four-level laser system has a LLL very close to ground, it is possible that the LLL will become thermally populated and so will behave, to some extent, as a three-level laser, where absorption (or re-absorption) of radiation at the lasing wavelength can occur.

While thermal population of the LLL is insignificant for the majority of four-level lasers (since many feature a LLL far above ground), some four-level lasers have a LLL close enough to ground such that, at operating temperatures, significant populations will be found:. Lasers in which the LLL has a "significant" thermal population at room temperature are termed *quasi-three-level lasers* and require modifications to mathematical models already developed.

EXAMPLE 5.1   ESTIMATING THE THERMAL POPULATION OF LLLS

The Nd:YAG operating at 1064nm is one of the most common solid-state lasers. The LLL is 0.262 eV above ground. Using the Boltzmann equation (Equation 1.2), the fractional population of the LLL may be estimated to be:

$$\frac{N}{N_0} = e^{\frac{-E}{kT}} = e^{\frac{-0.262eV \times 1.602 \times 10^{-19}}{1.38 \times 10^{-23} \times 300K}} = 3.98 \times 10^{-5}$$

So, at normal operating temperatures of about 300K, the thermal population of the LLL of the 1064nm transition for Nd:YAG is predicted to be 0.004% of the total number of lasing ions in the rod. Consider, as well, that an average YAG rod has a doping concentration of 1% or about $1.4*10^{20}$ neodymium ions per cm^3, and so for a YAG laser producing a gain (i.e. g_0) of 5m^{-1}, the number of ions involved in the inversion is:

$$\Delta N = \frac{g}{\sigma_0} = \frac{5}{2.8 \times 10^{-23}} = 1.79 \times 10^{23} \, m^{-3}$$

As a fraction of the doping concentration in the rod, this represents:

$$\frac{1.79 \times 10^{17} \, cm^{-3}}{1.4 \times 10^{20} \, cm^{-3}} = 0.00127$$

or about 0.13% of all ions in the rod are actually involved in the inversion that produces laser gain. This is over thirty times larger than the thermal population of the LLL and so clearly the thermal population of the LLL is insignificant to the lasing process for a 1064nm YAG laser (i.e. it is a "true" four-level laser).

Repeating the same calculations for the 946nm transition of YAG in which the LLL is 0.106eV above ground state, the thermal population is 1.65%—an enormous population that will surely affect the laser.

The reader is cautioned that this simple approach provides only a rudimentary estimate—a more detailed method which takes Stark splitting of the LLLs into account will be presented later in this chapter.

To define "significant" population consider the two transitions of the YAG laser shown in Figure 5.2, in which the LLL for the 1064nm transition is 0.262eV above ground, and the LLL of the 946nm transition 0.106eV above ground. As per Example 5.1, the thermal population of the LLL (at room temperature) was calculated to be 1.65% for the 946nm transition and 0.004% for the 1064nm transition. At 1064nm, then, the fraction of ions thermally populating the LLL is far less than the fraction involved in the normal inversion (i.e. $N_{LLL\text{-}Thermal} \ll \Delta N$), and so thermal population does not significantly affect the threshold equation nor the minimum pump power required to achieve lasing: This defines a "true" four-level laser. The thermal population of the LLL of the 946nm transition is a different story—it is an enormous population that will certainly affect the minimum pump power required. This defines a quasi-three-level laser.

This is a very simplified view of the problem, since real solid-state materials have LLLs split by the Stark effect, which ultimately results in a lower population at the specific level that serves as the LLL. We will examine this effect later in this chapter, but for now, the approach of using the Boltzmann equation to predict the population of the LLL allows a simple determination of the nature of a quantum system, i.e. whether it is truly four-level or quasi-three-level.

Whether a laser is defined as four-level or quasi-three-level also depends on the temperature employed. Most quasi-three-level lasers can be operated as true four-level lasers by reducing the operating temperature, which in turn reduces the thermal population of the LLL. Cryogenic cooling (for example, operating a laser immersed

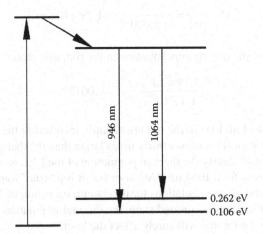

FIGURE 5.2   Transitions in the YAG laser.

in liquid-nitrogen at 77K) will allow almost any solid-state laser to operate as a four-level laser. It is, however, inconvenient.

So, where does this heat originate? The presence of heat in a laser crystal is inevitable, as we shall see in the next section.

5.3   QUANTUM DEFECT HEATING

The presence of heat in a four-level laser is unavoidable and originates from what is termed *quantum defect*. In any laser system (at least common ones), a relatively large pump photon is required to produce a smaller lasing photon. The difference in energy between these two photons ($hv_{PUMP} - hv_{LASER}$) is termed the *quantum defect* and results in heat that, in turn, causes thermal populations to build in the LLL. In some cases, quantum defect is expressed as a percentage of pump photon energy according to:

$$\left(1 - \frac{hv_{LASER}}{hv_{PUMP}}\right) \times 100\% \tag{5.1}$$

EXAMPLE 5.2   QUANTUM DEFECT CALCULATIONS

Consider a 1064nm YAG laser pumped by an 808nm semiconductor laser. Using Equation (5.1) the quantum defect is calculated to be:

$$\left(1 - \frac{1.87 \times 10^{-19}}{2.46 \times 10^{-19}}\right) \times 100\% = 24\%$$

So, 24% of the pump energy is wasted in the form of heat and, conversely, 76% of pump energy is converted into laser output. This sets the upper limit on the efficiency of this laser (which is quite high). With a large pump power a large amount of heat (24% of the total pump power) will be produced within the amplifier. If this were a quasi-three-level transition this would certainly cause large re-absorption loss, but since the LLL for this transition is far above ground state no significant population will result.

Quantum defect presents an ultimate limit on the efficiency of a laser since even if every pump photon resulted in a laser photon, 100% efficiency could never be achieved. Furthermore, where a laser medium is quasi-three-level the ramifications of quantum defect are that there will always be thermal energy available to populate the LLL and so re-absorption cannot be completely suppressed.

A small quantum defect means potentially high efficiency, but it also means high thresholds for pump power.

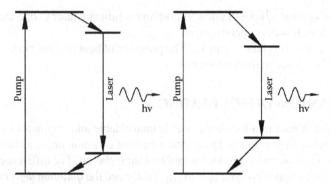

FIGURE 5.3 Quantum defect in a laser.

As depicted in Figure 5.3, quantum defect is a function of the relative positions of the quantum levels in the system, with the system on the left having a lower quantum defect than the system on the right.

It is inevitable that materials that have a low quantum defect also exhibit quasi-three-level behavior since a low quantum defect often dictates that the LLL be close to ground state (and hence will become thermally populated at "normal" temperatures). Quantum defect is hence a double-edged sword: On one hand, a small defect means the LLL is close to ground and so energy is not wasted and overall efficiency is high. On the other hand, having the LLL close to ground means it is easily populated thermally so that the LLL fills, reducing the available inversion (and, as we shall see later in this chapter, increasing the minimum pump power required to bring the laser to threshold).

Most "good" solid-state lasers have a LLL far enough above ground so that thermal populations are low but high enough that efficiency is still reasonable.

For some materials, quantum defect can be minimized by choosing a different pump wavelength. Consider, again, Nd:YAG oscillating at 946nm, with the key levels outlined in Figure 5.4. While normally pumped at 808nm, this same material can be pumped at

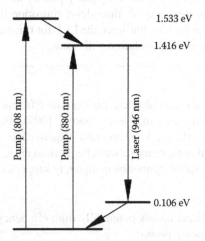

FIGURE 5.4 Alternative pumping wavelengths.

a longer wavelength of 880nm, which will result in a lower quantum defect and hence lower intrinsic heating. (It will still behave as a quasi-three-level material, but the amount of heat produced in the material will be reduced, as will the LLL thermal population.)

While the 1064nm transition of Nd:YAG is the most common, the 946nm transition outlined in this section is of interest (despite the fact that it is quasi-three-level) since doubling this wavelength results in a blue wavelength at 473nm, which is very close to an established line of the argon-ion laser, hence potentially offering a solid-state replacement for this gas laser.

Aside from the example of the 946nm transition of YAG, another quasi-three-level material of interest is Yb:YAG, which oscillates at 1031nm—this wavelength being of interest since the second harmonic is almost exactly that of the green argon-ion laser at 514.5nm, which has many established applications. This material has a particularly small quantum defect, which promises high efficiency.

Finally, the decision of pump wavelength depends on more than quantum defect, since it also depends on absorption of the pump wavelength by the amplifier material as well as cost of the laser diode. (Some pump diodes are more common and so are available at lower cost or at higher powers.)

5.4 THERMAL POPULATIONS AT THRESHOLD

The presence of thermal populations at the LLL can markedly increase the minimum pump power required to bring a laser to the threshold of oscillation. In order to model this increase, consider the most simplistic approach toward thermal populations by returning to the definition of gain from Equation (2.20):

$$g = \Delta N \times \sigma_0 = (N_{ULL} - N_{LLL})\sigma_0 \tag{5.2}$$

So, if threshold gain is used, ΔN is the threshold inversion. Where a significant LLL population exists (as it would in a quasi-three-level laser) this may be computed using the Boltzmann equation (Equation 1.2), and so the required ULL population may be computed (which is simply the inversion required for threshold plus the LLL thermal population).

An alternative, and more common, approach is to express the LLL as an absorption loss figure that may simply be added to the threshold gain figure, which already represents the loss due to the optical elements of the laser. This re-absorption loss is expressed as:

$$\gamma_{THERMAL} = \frac{\sigma_0 N_{LLL}}{Volume} \tag{5.3}$$

where σ_0 is the cross-section and N_{LLL} the population of the LLL as predicted by the Boltzmann equation. The resulting units are in "per unit length," the same units as

Quantum defect is not related to quantum efficiency (which is the fraction of pump photons that actually contribute to production of a laser photon).

"normal" absorption (γ), which originates due to optical losses such as attenuation and scattering in the amplifier.

Alternately, for a solid-state laser, the doping density is often known and so the population density of the LLL is computed. In this case,

$$\gamma_{THERMAL} = \sigma_0 N_{LLL} \tag{5.4}$$

where N_{LLL} is now the population density of the LLL in units of m^{-3}. The above form of re-absorption loss is common in literature such as research papers (and the reader is hence warned about the ambiguity of the term N_{LLL}, which can mean an absolute value or a density).

The threshold gain (i.e. the total gain the medium must produce in order to overcome all losses and bring the laser to oscillate) is then computed, at least for a simple laser as derived in Equation (2.5), as:

$$g_{th} = \gamma_{THERMAL} + \gamma + \frac{1}{2x}\ln\left(\frac{1}{R_1 R_2}\right) \tag{5.5}$$

For the purposes of calculating minimum pump power required, then, re-absorption loss can be modeled as a loss in the same manner as cavity optics. In the absence of significant quantities of cavity photons (small-signal), the loss will be approximately constant and dependent solely on the thermal population of the LLL as predicted by the Boltzmann equation. It is worth mentioning, as well, that loss due to thermal population of the LLL is saturable and that the approach outlined here applies only to a laser operating below or at threshold. Once oscillation is achieved (i.e. operation above threshold), a different approach is required, as outlined in the next section.

From what we have seen so far about real atomic systems, it would be naïve to expect that each energy level in a given atomic species was actually discrete. In reality, the lasing levels of most materials (especially solid-state laser materials such as YAG) are split by the Stark effect, which is an electrical effect analogous to the magnetic Zeeman effect, into several hyperfine levels.

Figure 5.5 shows a hypothetical energy level X split by the Stark effect into numerous levels labeled L_1 through L_n.

As per the method outlined by Fan and Byer,[1] the thermal population of the actual LLL employed by a transition may be computed as a fraction of the total level according to the simplified partition function:

$$f_L = \frac{g_x e^{\frac{-E_x}{kT}}}{\sum_i g_i e^{\frac{-E_i}{kT}}} \tag{5.6}$$

Stark splitting of atomic energy levels is the electrical analog to the magnetic Zeeman effect.

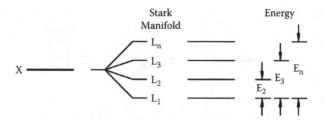

FIGURE 5.5   Stark splitting of an energy level (generalized).

where g_x and g_i are the degeneracies of the levels involved (degeneracy being the number of atomic levels at exactly the same energy), and E_x and E_i the energies of the individual levels referenced to the lowest level in the Stark manifold (in this case, the energy of level L_1, which is zero for this example).

This partition function computes the distribution of atomic populations—atoms thermally promoted to an atomic "level," which in this case consists of a host of fine levels closely spaced together that fill according to Maxwell-Boltzmann statistics.

When Equation (5.6) is expanded, the resulting Boltzmann occupation factor f_L for the L_n level is computed as:

$$f_L = \frac{g_n e^{\frac{-E_n}{kT}}}{g_1 e^{\frac{-E_1}{kT}} + g_2 e^{\frac{-E_2}{kT}} + \ldots + n e^{\frac{-E_n}{kT}}} \tag{5.7}$$

Since the manifold includes ground state, the actual population of the L_n level, as a fraction (i.e. $N_{\text{Level-}n}/N_0$), is simply the fraction f_L for that specific level. Multiplying this fraction by the atomic density will yield an answer in terms of re-absorption per unit length as follows:

$$\gamma_{THERMAL} = \sigma_0 (f_L N_0) \tag{5.8}$$

Finally, threshold gain may be computed by adding this re-absorption term to the gain threshold equation.

EXAMPLE 5.3   THE MINIMUM PUMP POWER REQUIRED FOR AN Nd:YAG LASER OPERATING AT 946nm

Previously (in Example 2.10), the threshold gain and minimum pump power for a 1064nm YAG laser were computed. We now modify this solution for the 946nm quasi-three-level transition by including re-absorption loss and the required change in the cross-section of the transition.

To determine the population of the LLL of the 946nm transition, one must take into account Stark splitting of the level where an applied electric

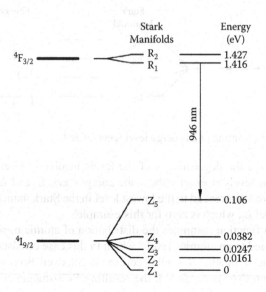

FIGURE 5.6   Stark levels of Nd:YAG.

field causes a level to split into many sub-levels. The LLL of the 946nm transition in Nd:YAG is normally described as a single level, denoted $^4I_{9/2}$, which is 0.106eV above ground state, but in reality it consists of five sub-levels, designated Z_1 through Z_5, as outlined in Figure 5.6. When thermal energy is taken into account, the total population of the LLL is split between these five levels and so the thermal population of the actual LLL is lower than expected.

The methodology outlined in this section will be used here with f_L computed for the 0.106eV energy level, which serves as the LLL for the 946nm transition. In the text, references to "atomic" population are interchangeable with "ionic" population since the actual lasing species is an ion (Nd^{3+}).

Expanding Equation (5.6) from this section, one can compute the resulting Boltzmann occupation factor for the Z_5 level as:

$$f_L = \frac{g_5 e^{\frac{-E_5}{kT}}}{g_1 e^{\frac{-E_1}{kT}} + g_2 e^{\frac{-E_2}{kT}} + g_3 e^{\frac{-E_3}{kT}} + g_4 e^{\frac{-E_4}{kT}} + g_5 e^{\frac{-E_5}{kT}}}$$

Following Equation (5.6), the fractional population of each Stark sub-level can be calculated (where YAG has five sub-levels). The LLL of the 946nm transition is the uppermost of these five and so n in the previous example is actually 5 for this transition. The degeneracies of all levels (g_1 through g_5 in the above equation) in Nd:YAG is 2. Assuming a normal operating temperature of 300K,

---

TABLE 5.1

Fractional Populations of Stark Levels

Stark Level	f_L
Z_5	0.0076
Z_4	0.1052
Z_3	0.1777
Z_2	0.247
Z_1	0.461

---

the resulting numerical solutions for the fractional population of each Stark level shown in Figure 5.6 are shown in Table 5.1.

These fractional populations, as required, sum to 1.

So, at 300K, a total of 0.76% of all atoms will thermally populate the LLL (designated Z_5) of the 946nm transition. This is a relatively large population for a four-level laser and one which will certainly increase the minimum pump power required for this laser. Contrast this to a "simplistic" approach outlined at the beginning of this section that neglects Stark splitting altogether, namely,

$$f_L = e^{\frac{-0.262eV \times 1.602 \times 10^{-19}}{1.38 \times 10^{23} \times 300K}} = 0.0165$$

The actual thermal population, then, is considerably less than predicted using a simplistic approach and so re-absorption loss is not as bad as might have been originally thought.

Knowing that the nominal doping density of Nd:YAG is about 1%, or $1.36*10^{20}cm^{-3}$, and that the cross-section of the 946nm transition is $5*10^{-20}cm^2$, the re-absorption loss term is then computed using Equation (5.4):

$$\gamma_{THERMAL} = \sigma_0(f_L N_0) = 5 \times 10^{-20}\,cm^2 \times 0.0076 \times 1.36 \times 10^{20}\,cm^{-3} = 0.0517cm^{-1}$$

or 5.17m^{-1}. In this case, $f_L N_0$ is the actual LLL population—with N_0 used since it is the population of the lowest level of the manifold. Since this manifold, like most other quasi-three-level lasers, includes the ground state at 0eV (and e^0 is 1) the LLL population is simply the fraction already calculated, multiplied by the number of lasing ions in the amplifier volume. It might be noted as well that the population density is used here directly to render an answer in units of "per unit length," the units usually used for attenuation or gain.

The minimum pump power is now computed in the manner described in Example 2.7 but with an increased threshold gain.

Assume a laser with the following parameters:
$R_{HR} = 100\%$
$R_{OC} = 99\%$
Active volume = 0.5mm diameter by 1mm length
Attenuation in YAG = $0.1m^{-1}$

Assuming this is a simple laser and that the lengths over which attenuation and gain occur are equal, Equation (2.5) may be used to compute the threshold gain (without re-absorption loss) for this laser to be $5.125m^{-1}$. If re-absorption loss is included (as computed in this example), the threshold gain is found to be $10.295m^{-1}$.

As per the methodology of Section 2.7, using the effective ULL lifetime of 230μs and a cross-section for the 946nm transition of $5*10^{-24}m^2$ (both accepted values found in literature) the rate of decay from the ULL, and then the minimum pump power required to bring the laser to threshold is then found using Equation (2.30) to be:

$$P_{MIN} = \frac{g_{th}h\nu_{PUMP}}{\sigma\tau}V = \frac{10.295m^{-1}hc}{808\times10^{-9}m(230\times10^{-6}s)(5\times10^{-24}m^2)}$$

$$(0.001m \times 0.00025m^2\pi) = 0.432W$$

So, while the calculated minimum pump power is 212mW in the absence of thermal re-absorption, the presence of thermal populations at the LLL more than doubles this figure.

Cooling the laser to 250K (which can be accomplished by a small and inexpensive Peltier thermoelectric cooler) will lower the fraction of atoms (f_L) in the LLL to 0.00369, which will lower the thermal re-absorption loss to $2.51m^{-1}$ and hence the pump power required accordingly. Many quasi-three-level lasers are actively cooled for this reason.

One might note that the YAG crystal in this example is quite small. There are a number of reasons for this including the ability to efficiently end-pump the laser, which ultimately increases the gain in a small active volume resulting in a large photon flux. Because re-absorption loss is saturable, a large photon flux is desirable for efficiency. This is considered in Section 5.5.

EXAMPLE 5.4   COMPUTING FRACTIONAL POPULATIONS

To further illustrate how fractional populations are computed, let's compute the relative populations of the ULL for YAG (since the actual energy of the lowest level in this Stark manifold is not zero) as follows.

Referring to Figure 5.6, the lowest level in the ULL Stark manifold is the R_1 level, which will be referenced as $E_1 = 0$. The energy of level R_2 is hence the difference between the two levels and so $E_2 = 0.011\text{eV}$. The population of the R_1 level is hence computed using a Boltzmann distribution as:

$$e^{\frac{-0}{1.38\times10^{23}\times300K}} = 1$$

As expected, the relative population of this level is 1. The population of the R_2 level is similarly computed:

$$e^{\frac{-0.011eV\times1.602\times10^{-19}J/eV}{1.38\times10^{23}\times300K}} = 0.653$$

The resulting fractional population for the R_1 level is:

$$\frac{1}{(1+0.653)} = 0.605$$

and for the R_2 level,

$$\frac{0.653}{(1+0.653)} = 0.395$$

So, at 300K, 60.5% of the atomic population will be found at the R_1 level (from which the 946nm transition originates). This fraction, along with the fractional population of the LLL computed in the first example, is required in Section 5.5 when considering the situation of a quasi-three-level laser operating above threshold (i.e. oscillating).

5.5   THERMAL POPULATIONS IN AN OPERATING LASER

Re-absorption loss affects the minimum pump power required to bring a laser to threshold, and in a quasi-three-level laser the effect can be enormous. A quasi-three-level laser is not completely limited, however, since above threshold, re-absorption loss is saturable and so eventually becomes very small, often insignificant, compared to other losses in the system. As per the methodology outlined by Fan and Byer,[1] re-absorption loss as a function of cavity radiation intensity (above threshold) is:

$$\delta_{THERMAL} = \frac{\sigma\Delta N_0 xI_{SAT}}{I}\ln\left(1+\frac{2I}{I_{SAT}}\right) \tag{5.9}$$

where $\delta_{THERMAL}$ is the re-absorption loss expressed as a dimensionless quantity—in other words, the "per unit length" loss (normally in units of m^{-1}) multiplied by the round-trip length of $2x$. ΔN_0 is the inversion in volumetric (m^{-3}) units and I_{SAT} is the saturation intensity which, in the simplest form, is the same as that already derived:

$$I_{SAT} = \frac{h\upsilon}{\sigma\tau} \tag{5.10}$$

As per the previous section, in a solid-state laser where Stark level splitting is taken into account this quantity becomes

$$I_{SAT} = \frac{h\upsilon}{(f_L + f_U)\sigma\tau} \tag{5.11}$$

As before, f_L is the fraction of atoms occupying the LLL and f_U is the fraction occupying the ULL since it, like the LLL, is split into multiple levels—in this case, two Stark levels. I_{SAT}, then, takes into account the population of the actual lasing levels involved.

At low intensities, where $I \ll I_{SAT}$, an approximation may be applied as follows:

$$\frac{I_{SAT}}{I} \ln\left(1 + \frac{2I}{I_{SAT}}\right) \approx 2 \tag{5.12}$$

And so, Equation (5.9) simplifies to:

$$\delta_{THERMAL} = 2\sigma\Delta Nx \tag{5.13}$$

or, expressed as a "per unit length" quantity by dividing by the round-trip length of $2x$,

$$\gamma_{THERMAL} = \sigma\Delta N \tag{5.14}$$

This is exactly the equation (5.4) that was derived earlier when dealing with threshold (assuming the "inversion" is now really the opposite of that normally considered—a large population at the LLL and so a loss)!

Now, at high intensities, the value of $\delta_{THERMAL}$ decreases since the term (I_{SAT}/I) becomes quite small and so eventually becomes insignificant beside other losses in the system (including absorption in the medium and the loss due to the cavity optics). In terms of slope efficiency of the laser, when cavity radiation greatly exceeds saturation intensity, the slope efficiency of a quasi-three-level laser approaches that of a four-level laser at large intensities.

> In the notation used here, γ is a distributed loss (in units of m^{-1} or cm^{-1}) while δ is a dimensionless loss ("per round trip").

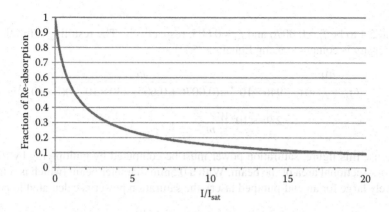

FIGURE 5.7   Re-absorption loss as a function of intensity.

As described in Equation (5.9), the saturable term is:

$$\frac{x\, I_{SAT}}{I} \ln\left(1 + \frac{2I}{I_{SAT}}\right) \tag{5.15}$$

which will vary from $2x$ when $I \ll I_{SAT}$ to essentially zero when $I \gg I_{SAT}$. The value of this term, divided by $2x$ to represent the total round-trip loss, is plotted as a function of the ratio of intra-cavity intensity to saturation intensity on Figure 5.7.

Thermal re-absorption, then, is a function of intra-cavity power, or rather the ratio of this power to the saturation power. Since saturation power is a function of beam diameter, smaller beam diameters mean smaller saturation intensities and hence higher saturation of re-absorption loss.[2] For this reason, end-pumping is often used with quasi-three-level lasers and pump beams are tightly focused to keep intensity very high in the active volume used in the amplifier.

EXAMPLE 5.5   PUMPING A 946nm Nd:YAG LASER

The design of quasi-three-level lasers is influenced greatly by the relations in this section. Consider first the saturation intensity of Nd:YAG operating on the 946nm transition. This intensity is computed according to Equation (5.11) with the upper and lower population fractions computed previously in Examples 5.3

Lasers using quasi-three-level amplifiers are often end-pumped to increase intra-cavity flux and hence reduce losses to re-absorption.

and 5.4 to be $f_L = 0.0076$ and $f_U = 0.605$, respectively. The resulting saturation intensity is computed using Equation (5.11) as:

$$I_{SAT} = \frac{h\upsilon}{(f_L + f_U)\sigma\tau} = \frac{hc}{946 \times 10^{-9} m(0.0076 + 0.605)(230 \times 10^{-6} s)(5 \times 10^{-24} m^2)}$$

$$= 2.98 \times 10^8 \, W\big/_{m^2}$$

To use this figure, saturation power must be computed by multiplying by the cross-sectional area of the beam. With a 0.5mm diameter beam (which is relatively large for an end-pumped laser), the saturation power is calculated to be:

$$P_{SAT} = I_{SAT}\pi r^2 = 58.6W$$

Assuming a 1% transmitting OC, this would correspond to an output power of 586mW when $I = I_{SAT}$—a level far in excess of what would be considered "reasonable" for a small laser of this type, and even at this intensity, the ratio $I = I_{SAT}$ means (according to Figure 5.7) a significant loss exists in the system due to re-absorption.

Now consider a beam that is focused to a much tighter diameter of 0.1mm. The saturation power in this case is found to be 2.34W. Again, assuming a 1% transmitting OC, this would imply an output power of 23.4mW when $I = I_{SAT}$. Normally, one might expect a laser of these dimensions to have an output power around 100mW, and so the actual ratio I/I_{SAT} is approximately 4 with the resulting re-absorption loss (from Figure 5.7) being reduced to about one-quarter of the original value with the larger beam diameter.

The ramifications to laser design are that where a transition has a small cross-section, and hence a resulting large saturation intensity, end-pumping is often used to decrease the size of the cross-sectional area of the beam mode within the amplifier. This decreases the saturation power such that favorable ratios (according to Figure 5.7) can be maintained and as such re-absorption loss becomes minimized.

The disadvantage of end-pumping is limited output as a result of limited power available from a single pump diode. While a single pump-diode is limited, practically, to a few watts, side-pumping allows use of high-power arrays that can generate hundreds of watts of power. Returning to the expression for output power (Equation 3.14), output power is proportional to saturation power, so large beam diameters, or side-pumping, offer the potential for higher output powers, but at a cost of higher threshold gain due to increased thermal re-absorption, as well as a large re-absorption loss unless a large intra-cavity power is allowed to build.

The more common 1064nm YAG transition does not present these same design challenges since the saturation intensity is almost ten times lower than that of the 946nm transition and the lower level does not thermally populate. This transition is a "true" four-level transition.

Note that for complete accuracy, one must average over the intensity of the Gaussian beam (assuming the laser was operating in a TEM_{00} mode—otherwise an intensity distribution reflective of the mode used must be employed). Refer back to Section 3.6, where the issue of beam profile was examined.

5.6 THERMAL EFFECTS ON LASER DIODES (WAVELENGTH)

Laser diodes are particularly sensitive to thermal effects, but in a different manner to, say, a gas or solid-state laser. The most pronounced thermal characteristic of diodes is the tendency of the wavelength to drift with temperature—with many diodes the output wavelength will shift between 0.25 and 0.35nm per degree C of temperature change.

It is well known that temperature has an effect on all semiconductor diode junctions—silicon diodes, for example, are often used as temperature sensors by measuring the bandgap voltage of the device, which is a function of temperature (as temperature increases, bandgap voltage decreases)—so it is not surprising that the peak wavelength of the emission of a diode laser also depends on temperature. Typically, a laser diode drifts approximately 0.25nm/°C (for a red or near-IR diode). This characteristic may be used to advantage to temperature-tune a diode to a specific required wavelength. (This is often done when a diode is used as a pump source for a solid-state laser.)

Predicting these effects theoretically can be complex, but experimental observations (usually outlined by the manufacturer on a device datasheet, but simple enough to accomplish with the help of an OSA) yield a simple relationship between wavelength and temperature. The datasheet for a typical laser diode used for pumping of small DPSS vanadate lasers is shown in Figure 5.8. This laser has a nominal output of 200mW at 808nm although, like any laser diode, the output wavelength is dependent on temperature—this is evident from the graph on the second page of the datasheet (expanded in Figure 5.9). Between 20°C and 30°C, the wavelength of the device varies 0.22nm/°C spanning a range of 807nm to 809.2nm and between 30°C and 40°C the wavelength of the device varies 0.26nm/°C until 811.8nm is reached. While the diode can operate outside this range, these are the only ranges of wavelengths of concern for pumping vanadate, which has a narrow absorption peak at 808.8nm.

Output from many semiconductor lasers is not at a single wavelength, but rather spread across a range of wavelengths since the ULL and LLL are in reality energy bands with finite width. Populations of electrons can assume a large number of energies within these bands leading to a wide gain curve, typically from 1nm to 3nm in width (Full-Width Half-Maximum [FWHM]). As the temperature of the device changes, the entire gain curve shifts.

Of course, like other lasers the presence of a resonant cavity ensures the output consists of multiple longitudinal modes, each at a slightly different frequency. Given the small dimensions of the optical cavity of a diode (often 300μm to 500μm in length), modes are spaced over 0.2nm apart and so can easily be observed with an OSA with modest resolution (and given that the spectral width of a diode is commonly 1nm to 3nm, the output from such a diode may consist of fifteen modes or more).

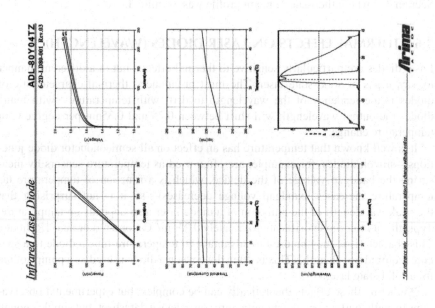

FIGURE 5.8 Laser diode datasheet. (Courtesy Arima Lasers, Inc.)

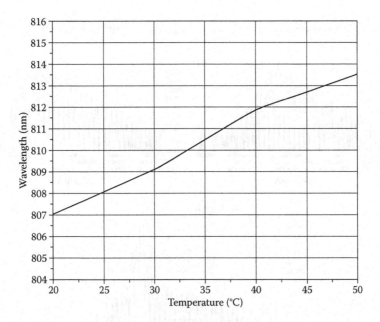

FIGURE 5.9 Dependency of wavelength on temperature.

The actual shape of the diode output depends on a number of factors such as operating current. At low currents the output is quite broad, consisting of many longitudinal modes that take the general shape of the gain curve (which is more or less Gaussian). At all points where the gain exceeds losses in the device, output will appear. As current increases, central modes tend to become dominant and spectral width tends to narrow. (The mechanism by which the spectrum is broadened in a diode is homogeneous.) While some laser diodes are designed to operate on a single longitudinal mode (often called SLM diodes), the output of essentially all diode lasers will narrow at high drive currents, often with one single mode becoming dominant to the point where the output appears single mode regardless. This is demonstrated in Figure 5.10 where the output from a common, inexpensive laser diode is seen at various drive currents. As drive current was increased, the central mode increased in power until finally becoming dominant. In addition, the shift toward longer wavelengths is evident.

To complicate matters, consider as well that the characteristics of the cavity can also shift with temperature. The absolute wavelengths at which resonance occurs depends on the cavity spacing (in a laser diode, the length of the actual device), but the index of refraction of the semiconductor material itself changes with temperature and so, effectively, the cavity "lengthens" optically as temperature increases. This

SLM diode lasers tend to have lower output powers than multi-mode diodes and so are uncommon for pumping solid-state lasers.

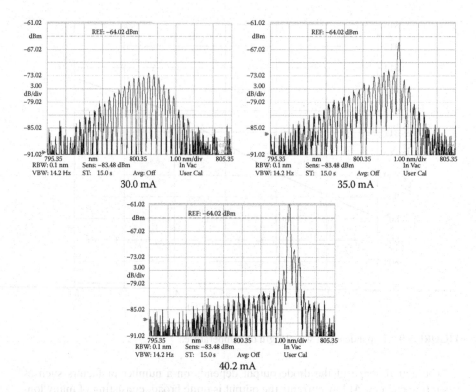

FIGURE 5.10 Laser diode output at various drive currents.

can cause a shift in observed wavelength as well (coupled with the shift due to the movement of the gain peak).

Where a laser supports only a single longitudinal mode (or single frequency), or even when a single dominant mode appears in the output of a "regular" diode laser, the effects of temperature change can lead to a rather startling effect known as mode hopping, in which the output wavelength changes in sudden jumps as the dominant mode shifts. Changes in temperature cause the wavelength at which maximum gain occurs to move from one cavity mode to the next. Unlike a multi-mode laser, where oscillation on many simultaneous wavelengths leads to an averaging so that the center wavelength appears to slowly drift, wavelength of the single mode makes abrupt jumps. This can be seen in Figure 5.11, which shows the wavelength of a standard laser diode changing in steps while mode hopping. (In this case, the diode was driven to maximum current which caused a single dominant mode to appear.) It is also interesting to note that during this experiment some "jumps" were observed to move two modes over instead of to the next adjacent mode as expected.

Short lasers, with FSR peaks spread far apart, can help eliminate mode hopping since small changes in temperature, with corresponding small changes in the wavelength of maximum gain, will not result in that same gain peak moving to the next resonant mode. Use of a Distributed Bragg Grating (DBG) as one cavity optic can also help avoid mode hops, although temperature changes can affect the center

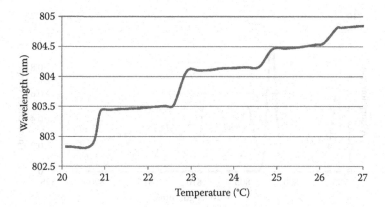

FIGURE 5.11   Mode-hopping in the output of a laser diode.

wavelength of such reflector structures as well. The effect is not as large as the afore-mentioned shift of the center wavelength of the gain medium and so the wavelength of such diodes drifts far less over a given temperature change than a "normal" laser diode using two plane broadband reflectors.

Unfortunately, it is not always easy to distinguish between single- and multi-mode lasing since many diodes will begin oscillation in multi-mode and as current increases, will become single mode.

5.7   MODELING THE EFFECTS OF TEMPERATURE ON LASER DIODES (WAVELENGTH)

Diode lasers are often used to pump small solid-state lasers. By "tuning" the output of the laser diode to match the absorption peak of the amplifier material, an extraor-dinarily efficient laser can be built. Nd:YAG, for example, will absorb only a few percent of pump radiation from an arc or flashlamp source—the overall wall-plug efficiency will be lower than this, given the efficiency of other components in the system—while diode pumping allows wall-plug efficiencies of over 50%!

Of course, the wavelength of the diode can drift, which now raises the question, "How would the output power of a DPSS laser change as temperature of the pump diode changes?" Assuming a "normal" laser diode is used, the output from the diode is not produced at a single, infinitely narrow wavelength, so one must consider how well all wavelengths of the diode spectrum "line up" with the corresponding absorp-tion peaks of the solid-state laser medium. A convenient technique to accomplish this

Convolution is a technique commonly used in Digital Signal Processing (DSP). For this purpose, most DSP processor chips feature a MAC (Multiply and Accumulate) instruction.

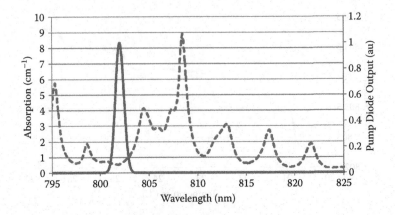

FIGURE 5.12   Pump diode output and YAG absorption.

is *convolution*, a mathematical technique where the "overlap" between two functions can be analyzed to produce a third function. Convolution is an important Digital Signal Processing (DSP) technique and is commonly used for filtering signals.

Consider the output from a laser diode (with a 1nm FWHM) along with the absorption of Nd:YAG, as shown in Figure 5.12. The center wavelength of the diode is shown at 802nm in the figure and clearly, at this wavelength, the absorption of the laser diode output will be poor. (In fact, unless the pump diode is quite powerful, it will likely not oscillate at all.) It is fairly obvious from the figure that for efficiency the narrow output spectrum of the diode must be shifted to match the large (but narrow) absorption peak of YAG located around 808.4nm. At this non-optimal wavelength of 802nm, the actual absorption of pump power from the laser diode is considered in Figure 5.13, for which we would need to calculate how well the output from the diode laser "overlaps" the absorption of

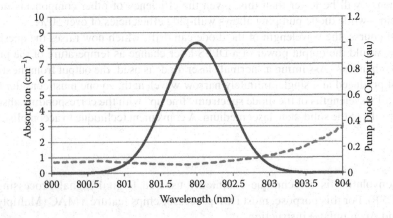

FIGURE 5.13   Pump diode output and YAG absorption expanded.

Nd:YAG. Since emission from the diode occurs across a range of wavelengths with various intensities, these wavelengths are absorbed to varying degrees and so convolution is used to determine the amount of total absorption of power from the pump diode.

The convolution technique is one of "sliding" the diode output curve along the absorption curve of the laser medium and multiplying the intensity at discrete wavelengths. It is a technique well suited to numerical (algebraic) solution (as is often done with DSP techniques). In Example 5.6, a spreadsheet is used to illustrate the process and produce a final analysis outlining the expected DPSS output vs. diode temperature.

If the two functions were known—call them $p(\lambda)$ for the output of the pump diode as a function of wavelength and $a(\lambda)$ for the absorption of the solid-state laser medium—the convolution of these (denoted as $p*a$ where $*$ is a star) is:

$$(p*a)(\lambda) = \int_{-\infty}^{\infty} p(x)a(\lambda - x)dx \qquad (5.16)$$

While the output from the pump laser diode can be modeled reasonably accurately as a Gaussian curve, the absorption of real materials is an irregular curve and so an algebraic approach is often employed in which discrete wavelengths are used as follows:

$$(p*a)(\lambda) = \sum_{x=-\infty}^{\infty} p(x)a(\lambda - x) \qquad (5.17)$$

Or, because of the commutative property,

$$(p*a)(\lambda) = \sum_{x=-\infty}^{\infty} p(\lambda - x)a(x) \qquad (5.18)$$

To illustrate this, consider again Figure 5.13, in which the pump diode emits a range of wavelengths centered around 802nm. The overall absorbed energy can be specified as:

$$p(800nm) \times a(800nm) + p(800.2nm) \times a(800.2nm) + p(800.4nm) \times a(800.4nm) + \cdots$$

$$(5.19)$$

where a discrete interval for wavelength of 0.2nm was chosen arbitrarily in this specific example. Often this interval is chosen by the resolution of the data available (especially of data originated from lab observations using, for example, an OSA to measure the output of a laser diode).

When complete, this operation will yield a numeric answer representing the total absorption of pump laser power with the diode operating at a center wavelength of 802nm (and so, at a specific temperature to yield that emission). The convolution operation is now repeated with the wavelength of the diode shifted (presumably 0.2nm, in this case), yielding a number representing the absorption with the pump diode centered at 802.2nm. This collection of points (pump diode center wavelength vs. total absorption at that wavelength) can be plotted to reveal the expected effect of the shift of the pump diode on the output of the DPSS laser.

The output spectrum of the laser diode is required for the model, both the general shape of the spectrum as well as the center wavelength as a function of temperature (since ultimately we are trying to determine the optimal temperature at which to operate the pump diode). The shape of the spectrum (i.e. intensity as a function of wavelength) may be researched from a datasheet (if this graph is available for the specific diode chosen), assumed to be a Gaussian profile, or determined experimentally using an OSA. Assuming this is not available from the datasheet, the next simplest method is to assume a Gaussian profile. Assuming the FWHM of the diode is known, a suitable profile may be generated by:

$$I(\lambda) = e^{\frac{-(\lambda - \lambda_0)^2}{2\sigma^2}} \tag{5.20}$$

where λ_0 is the center wavelength and λ is the variance that is related to the FWHM of the spectrum by:

$$\sigma = \frac{FWHM}{2.35} \tag{5.21}$$

As a comparison, the output spectrum profile (envelope, without longitudinal modes) of a real laser diode is graphed along with a Gaussian model of that same diode in Figure 5.14. The Gaussian profile was based solely on the observed spectral width (FWHM) for this diode. The actual diode, seen as a solid line in the figure, produces a somewhat narrower curve than the Gaussian profile as well as one that is skewed

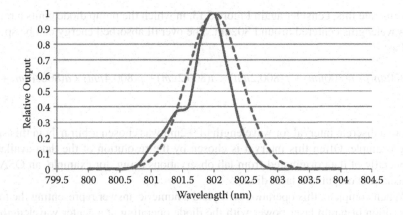

FIGURE 5.14   Actual laser output (solid line) vs. Gaussian profile.

toward shorter wavelengths. Where a diode has a multimode output spread over a wide bandwidth, this skewing towards shorter wavelengths is often observed. (It disappears when a single dominant mode is produced.)

The dependence of wavelength on device temperature is almost always specified in a diode datasheet in the form of a "wavelength vs. temperature" graph (as outlined in Figure 5.8). If required, this, too, could be determined experimentally by varying temperature of the device and observing the output, again with an OSA.

The other required parameter is the absorption spectrum of the laser medium. This, too, may be researched since the absorption curve for many common materials is well known and has been published, usually by the manufacturer of such materials. It may also be determined experimentally using a tunable laser such as a Ti:sapphire and measuring the absorption directly. For a numerical convolution, data for both curves (diode output as a function of wavelength and absorption of the laser medium as a function of wavelength) must be available at the same resolution—in this example, every 0.2nm.

Armed with data for each curve, one can easily implement a convolution algorithm with a spreadsheet or by writing a simple program.

EXAMPLE 5.6   PREDICTING THE EFFECT OF DIODE WAVELENGTH SHIFT ON VANADATE

Consider a practical example in which the effect of diode wavelength shift is predicted on vanadate ($Nd:YVO_4$). Vanadate is a material commonly used for small DPSS lasers. It is very efficient and features broader absorption peaks than does Nd:YAG, so it is less sensitive to diode wavelength shift than is Nd:YAG. (Compare the absorption curve for vanadate in Figure 5.15 to YAG in Figure 5.12 to get an appreciation for how sharp the absorption peaks for YAG really are. This places tight demands on the temperature control system for diodes pumping a YAG.)

A convolution model can be easily set up using a spreadsheet. Figure 5.16 shows such a spreadsheet along with an overlay describing the mathematical calculations involved and the diode output, which in this case is generated mathematically.

In the example shown, the predicted diode output (called the *kernel*, and shown in column D accompanied by a graphical curve depicting the output) is multiplied by absorption figures for wavelengths between 800nm and 804nm in column B.

Absorption values can be derived from experimental data or, for example, by digitizing absorption data provided by the manufacturer. For an experimental determination of the absorption curve, a tunable titanium-sapphire laser with a birefringent tuner was used as an optical pump source for a vanadate crystal. The laser features a narrow spectral width (of under 0.1nm), making it ideal for such determinations. Unfortunately, at regions of very low absorption (e.g. around 800nm), this method is limited by the output power available

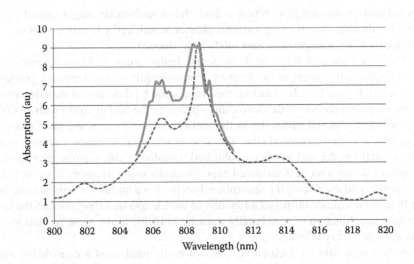

FIGURE 5.15 Vanadate (Nd:YVO$_4$) absorption.

	A	B	C	D	E	F	G
3	Vanadate Characteristic			Kernel		Output	
4	Wavelength (nm)	Absorption (cm-1)		Diode output			
5	800	1.2	⇨X⇨	2.97235E-08	⇨		
6	800.2	1.2	⇨X⇨	8.00223E-07	⇨		
7	800.4	1.22	⇨X⇨	1.52329E-05	⇨	Σ	
8	800.6	1.26	⇨X⇨	0.000205031	⇨		
9	800.8	1.36	⇨X⇨	0.001951264	⇨		
10	801	1.48	.	0.013130309	.		
11	801.2	1.62	.	0.06247352			
12	801.4	1.78	.	0.210173998	.		
13	801.6	1.9		0.499947031			
14	801.8	1.94		0.840874143			
15	802	1.92		1		7.873695	
16	802.2	1.82		0.840874143		7.713566	
17	802.4	1.72		0.499947031		7.517584	
18	802.6	1.68		0.210173998		7.38298	
19	802.8	1.68		0.06247352		7.36699	
20	803	1.72		0.013130309		7.483309	
21	803.2	1.78		0.001951264		7.723637	
22	803.4	1.86		0.000205031		8.078798	
23	803.6	1.96		1.52329E-05		8.543285	
24	803.8	2.12		8.00223E-07		9.106503	
25	804	2.26		2.97235E-08		9.746182	
26	804.2	2.47				10.43537	

FIGURE 5.16 A spreadsheet demonstrating convolution.

from such a laser. (The vanadate laser would not oscillate with a pump wavelength below 805nm due to the power levels required.) For comparison, the experimental results are graphed as a solid line along with data provided from a manufacturer of such crystals in Figure 5.15. Both methods correspond well, although the experimental data suggests higher absorption around 806nm than the manufacturer's data. Note that this graph (both theoretical and experimental results) are for a lightly doped vanadate crystal, and actual absorption can reach very large values with heavier doping concentrations (four times higher than Nd:YAG).

In either case, data points are available every 0.2nm (chosen for convenience based on the resolution of instrumentation available for this experiment).

Also required for this model is the diode intensity data (spectrum) provided at the same 0.2nm interval which can be sampled from a prospective device using an OSA or modeled using a Gaussian curve function as it is here. The actual pump diode used in this experiment was observed on an OSA and the FWHM determined from the resulting output (this was a built-in analytical function of the OSA instrument used so measurement of this parameter was trivial) was found to be 0.8nm. This procedure (modeling as a Gaussian profile using the observed FWHM) was used as opposed to providing observed intensity every 0.2nm.

In setting up the spreadsheet, there are 21 pairs of values in all (the number of data points for diode intensity data, now the size of the kernel), which are multiplied together and summed with the result appearing in cell F15—this sum corresponding to a diode center wavelength of 802nm. Note that the output from the convolution will always be shorter (i.e. have fewer points) than the input data (the absorption curve of vanadate in this case).

To accomplish the summing of these cells one could use a long formula in cell F15 of the form:

$$= (+B5*\$D\$5+B6*\$D\$6+B7*\$D\$7+ \ldots +B25*\$D\$25)$$

A simpler way to accomplish this, for spreadsheets that have this function, is the "SumProduct" function, which performs the same calculations:

$$= SUMPRODUCT(B5:B25,\$D\$5:\$D\$25)$$

Readers with experience using DSP hardware will recognize this function as being similar to the MAC (Multiply and Accumulate) instruction that is featured in essentially all DSP processors. (Indeed, the primary usage of a MAC instruction on a DSP processor is to execute convolution algorithms quickly.)

Note that both formulae references to the kernel in cells D5 to D25 are anchored so that they do not change, but references to the absorption figures will increment as the formula is copied to subsequent cells in column F. When

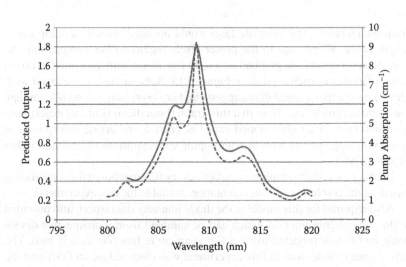

FIGURE 5.17 Output from the convolution model (solid line) along with absorption (dotted line).

the formula from cell F15 is copied to the next row (F16), the formula will become:

$$= \text{SUMPRODUCT}(B6:B26,\$D\$5:\$D\$25)$$

This same formula is copied to all cells in column F.

The output from the convolution model (column F) is shown in Figure 5.17, plotted against wavelength. As expected, sharp features are smoothed—the more broad the diode output, the less sensitive the output is to diode wavelength shift.

From the model provided, one can predict (a) the optimal temperature at which to operate the diode and (b) the effect of changes in that temperature. The model lends itself well to "What if?" scenarios such as, "What is the effect on output power if the temperature of the diode was allowed to change from x to y degrees?" It will also allow the determination of effects of various diode spectra (e.g. how sensitive the system would be to wavelength changes if a diode with a broad, or narrow, output spectrum is used).

Example 5.6 should nicely illustrate why practical, stabilized DPSS lasers require precise temperature control of not only the laser medium but also of the pump diode. In practical designs, device temperature is often chosen to correspond to a particular desired wavelength and is controlled tightly (within a fraction of a degree C) to ensure temperature drift does not occur. Smaller lasers often feature the diode mounted directly atop a thermoelectric cooler (TEC) module that utilizes the Peltier effect to pump heat away from the diode, as seen in Figure 5.18. These coolers are miniature heat pumps consisting of two ceramic plates interconnected by p- and

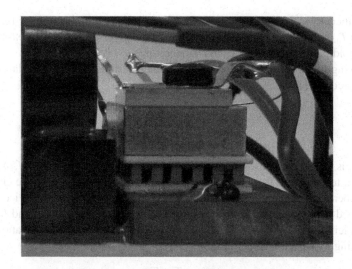

FIGURE 5.18   A diode atop a TEC (thermoelectric cooler) module.

n-type semiconductor materials (usually "pellets" of bismuth telluride doped to make them a semiconductor). When current flows through the device, heat is transferred from one side to the other with the amount of heat transferred a function of current.

A thermistor mounted on the same block as the diode (or other optical device) provides feedback to a PID temperature controller that adjusts current through the cooling module to keep the temperature constant. As the temperature rises (due to, for example, an increase in diode current), the PID controller will increase current to the Peltier cooler to provide more cooling power as required.

While Peltier coolers are the preferred method for small diodes (as well as other optical elements such as amplifiers and harmonic-generator crystals), with large diode arrays like those used to pump large solid-state lasers such as YAGs water cooling is often used where a high-capacity refrigeration system keeps the water at a constant temperature. Cooling water usually flows directly through the diode mounts.

5.8   THERMAL EFFECTS ON LASER DIODES (POWER AND THRESHOLD)

At a constant diode current, an increase in device temperature will result in a decrease in output power. Elevated temperatures will also result in an increase in threshold current. The sensitivity of laser diode threshold as a function of temperature

Characteristic temperature is a measure of how sensitive the threshold of a diode is to temperature change.

is quantified by a parameter called the *characteristic temperature* of the device (denoted T_0), which is empirically determined for a given device from measurements of threshold pump current at various temperatures.

Knowing the characteristic temperature, the threshold current for a laser device may be predicted by:

$$I_{TH} = I_0 e^{\frac{-\Delta T}{T_0}} \tag{5.22}$$

where I_0 is the known threshold current at an arbitrary temperature T, ΔT is the change in temperature from that same arbitrary temperature, and T_0 the characteristic temperature of the device. The characteristic temperature of a device is often determined experimentally by determining two threshold currents (I_1 and I_2) at two different temperatures (T_1 and T_2, respectively) and substituting into Equation (5.22), rearranging to solve for T_0 as follows:

$$T_0 = \frac{T_2 - T_1}{\ln\left(\dfrac{I_2}{I_1}\right)} \tag{5.23}$$

EXAMPLE 5.7   EXPERIMENTALLY DETERMINING CHARACTERISTIC TEMPERATURE

A laser diode was observed to have a threshold current of 52mA at a temperature of 25°C, and when heated to 50.0°C, the threshold current increased to 61 mA.

The characteristic temperature is found using Equation (5.23) to be:

$$T_0 = \frac{323.15K - 298.15K}{\ln\left(\dfrac{61mA}{52mA}\right)}$$

or 157K. A higher characteristic temperature represents a device that is less sensitive to temperature (i.e. more thermally stable).

Threshold current increases as a function of temperature, but once oscillation occurs slope efficiency is approximately the same at all temperatures. This can be seen in Figure 5.19, which shows the output power of a laser diode vs. current at temperature intervals of 10°C, ranging from 0°C to 40°C. The slope of the graph essentially remains constant with the threshold rising with temperature.

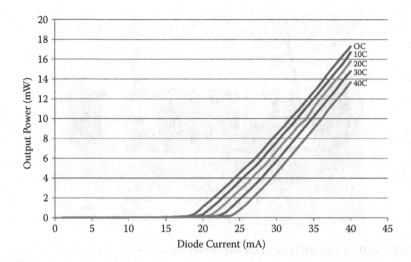

FIGURE 5.19   Laser output power as a function of temperature.

5.9 LOW POWER DPSS DESIGN

From the discussions in this chapter, it is evident that there are many parameters that affect the design of a laser. In this section, we consider the design of low power (under 10W) DPSS systems in which a laser diode pumps a solid-state amplifier.

First and foremost, the temperature of a laser diode affects the output wavelength of that diode, which, as evident from Section 5.7, can drastically affect the output of a laser—even a small drift in pump wavelength can result in a large swing in the output power of a DPSS laser. For this reason, temperature control of a pump diode is paramount for stability.

A secondary concern with a pump diode is power output, which decreases as temperature increases, as outlined in Section 5.8. And so, as far as the pump diode is concerned, there are two contradictory effects of temperature, namely:

1. As temperature increases, output power from the pump diode decreases (so output power decreases).
2. As temperature increases, wavelength from a pump diode shifts towards the YAG/vanadate absorption peak of 808.6nm (so output power increases).

Aside from the pump diode, thermal effects on the amplifier medium itself may be a concern. Depending on the amplifier medium (i.e. if it is a true four-level or a quasi-three-level medium), it may be advantageous to keep the solid-state material operating at as low a temperature as possible. In a quasi-three-level medium, elevated operating temperatures result in increased thresholds that, even when exceeded, require a highly concentrated beam diameter to saturate losses to keep efficiency high. With that in mind, whether the material is cooled or not, a high intra-cavity photon

IR Filter   Vanadate   Pump focussing Pump
                   /SHG   Optics   Diode

FIGURE 5.20   DPSS with two TECs (optics only shown).

intensity should be maintained in order to reduce re-absorption losses. With these materials, then, end-pumping is often used since this allows the pump beam to be focused tightly inside the material (and so the resulting laser mode is also small).

The most flexible design, from an optimization standpoint, would incorporate separate TEC controls for the pump diode and the amplifier (i.e. separate thermo-electric coolers, each with independent settings). This arrangement often results in a large physical separation between the pump diode and the amplifier allowing (or mandating) optics to focus and re-shape the pump beam, which exits the diode in a highly divergent beam with an elliptical shape. Figure 5.20 shows the optical train for such a DPSS laser in which the pump diode and the amplifier are mounted on separate TEC modules for independent temperature control. The laser in the photo shows the optical elements only. Wiring between elements such as the TECs, temperature sensors, laser diode, and photo sensor has been removed for clarity. (The only wires visible are those protruding from the Peltier coolers under the pump diode on the right and the amplifier on the left.) The amplifier in this design is a complete element with HR and OC deposited directly onto the ends of the crystal. Pumping is accomplished directly through the HR, on the side toward the pump diode, which is 100% reflecting at the lasing wavelength of 1064nm but transmits the 808nm pump beam from the diode.

Not evident in the figure is the fact that the amplifier (shown as "Vanadate/SHG" in the figure) also incorporates a second-harmonic generator (or SHG, covered in Chapter 7). This assembly, in which the amplifier (vanadate, in this case) is bonded directly with the harmonic generator crystal, is a complete laser unto itself with the OC and HR mirrors deposited directly onto the crystal faces. Such a hybrid crystal is seen in Figure 5.21 with a millimeter ruler for size comparison. The vanadate amplifier crystal is 1mm in length and is on the right side of the assembly. Careful examination will reveal a round dielectric HR mirror coating deposited directly onto the face of this crystal. The larger crystal on the left side is the harmonic generator (KTP in this case) with an OC mirror (not visible in the photo) deposited onto the

FIGURE 5.21 Hybrid laser/SHG crystal.

outside face of that crystal as well. The harmonic generator, then, is located inside the cavity of the main solid-state laser (i.e. placed between the vanadate amplifier and the OC) and converts infrared radiation into visible light—in this specific case, converting the 1064nm infrared of the vanadate laser into green light at 532nm. Without this harmonic generator, the laser would produce only IR output at 1064nm.

Harmonic generator crystals are very sensitive to temperature, and in this particular design are of far more concern than the temperature of the amplifier. While an increase in temperature of a few degrees will hardly make a difference in the efficiency of the vanadate (since it is a four-level material and thermal population of the LLL is very small at operating temperature), it will drastically affect the ability of the harmonic generator to produce green radiation. (A difference of only a few degrees can cause the output of green light to fall to half of the maximum value.) The temperature of the diode, then, is optimized to produce a peak output wavelength to match the absorption peak of vanadate, and the temperature of the amplifier/second harmonic generator hybrid crystal is optimized for second harmonic production (in what is called *phase matching*, covered in Chapter 7). The arrangement shown, with one TEC for the pump diode and a second for the amplifier/harmonic generator crystal, is not the only arrangement possible. Some manufacturers utilize one TEC for the pump diode and amplifier together and a second for the harmonic generator alone. Regardless, the temperature of the amplifier is usually not an overtly critical parameter (especially if the amplifier is four-level and not quasi-three-level) as opposed to the large losses that result when the wavelength of the pump diode deviates from the absorption peak of the amplifier.

One might wonder if the decrease in the output power of the diode due to temperature increase (as illustrated in Section 5.8) is a factor. Unfortunately, while cooling a laser diode will result in lower threshold currents and higher output power at the same drive current, the effect of wavelength shift is an overriding concern. For example, an 800mW, 808nm pump diode was tested at various temperatures with the result being an output power drop of less than 10% over the range 10°C to 40°C.

With the temperature shifting by the same amount, an average wavelength shift of 0.26nm/C was observed for a total of 7.8nm. A drift in output wavelength of 7.8nm in the pump wavelength will certainly affect the output power of a DPSS considerably more than a 10% drop in overall pump power!

The ability to control diode and crystal temperatures separately allows optimization of many parameters but precludes manufacturing of a compact system. The necessity for optics to focus pump radiation on the crystal and the need for two separate TEC controllers can make the system become quite complex and cumbersome. For this reason, while larger laser systems (e.g. laboratory lasers) often feature separate temperature controls for all elements, many compact (and simple) lasers often feature only a single TEC for all elements, with the pump diode and the amplifier operating at the same temperature. In forming a compact laser, the diode is usually placed in close proximity to the amplifier, sometimes (but not always) foregoing optics between the two elements completely. Where a harmonic generator is required (i.e. for visible laser output), a hybrid amplifier/harmonic generator crystal is used as described previously. This hybrid crystal is then attached, ultimately, to the same mount as the pump diode so all elements operate at the same temperature. Clearly, in this arrangement tradeoffs are required—the temperature is set such that the diode produces the optimal wavelength for pumping, the temperature of the amplifier is set low to minimize re-absorption loss, or the temperature is set for efficient harmonic production. In such an arrangement, pump diodes are often chosen so that their emission wavelength aligns with the peak absorption of the amplifier at approximately the same temperature that the harmonic generator achieves phase matching (and so acts as an efficient converted). Assuming the material is not operating on a quasi-three-level transition, the temperature of the amplifier will not be a significant factor compared to the other factors. With careful matching of design parameters, deviation from the optimal temperature will result in both a decrease in the efficiency of the harmonic generator and a decrease in output power given that the wavelength of the pump diode will drift away from the peak absorption band for the amplifier. Couple these two factors together and one can see why accurate temperature control is especially important in a design such as this that uses a single cooling module (for stability of output power, at least).

A laser of this type is pictured in Figure 5.22, which shows the components of such a laser in the foreground and a complete system in the background in which all components are mounted in a brass tube. The components, from left to right, are the pump laser diode, a lens to focus the pump radiation mounted inside a metal disc, hybrid vanadate/harmonic generator crystal complete with integral cavity mirrors (of the same type as employed in the previous laser), infrared filter on the same mount as the hybrid crystal (the square glass affixed to the front of that element), and focusing lens for the 532nm output beam. While a lens is included in this design to focus pump radiation from the laser diode onto the crystal, some very compact and inexpensive designs (e.g. green laser pointers) omit even this element, in which case the amplifier is situated extremely close to the pump diode since the beam emerging from the diode is highly divergent. Another feature of this design is the IR filter, which removes both residual 808nm pump radiation and residual 1064nm vanadate radiation from the output since some IR will often pass through the OC. It has been

FIGURE 5.22   A compact DPSS.

found that many inexpensive green laser pointers omit this infrared filter, resulting in a safety hazard. While the green output may be well under 5mW (and hence the laser pointer falls into a class-IIIa laser product category), the residual infrared output can exceed 30mW on such a laser, which represents a significant danger!

5.10   SCALING DPSS LASERS TO HIGH POWERS

High-power DPSS lasers suffer from the same issues as low-power lasers, the most prominent of these being the requirement (for high efficiency) to "temperature tune" pump diodes, so accurate temperature control for pumping high-power DPSS lasers is the rule. In the case of high-power lasers, end-pumping is also a limitation since single-emitter diode lasers are not scalable to the power levels desired of hundreds of watts, so arrays of laser diodes are used. These arrays, which can produce power levels of hundreds or even thousands of watts, are generally water-cooled to maintain a constant temperature (and hence output wavelength) and to remove heat to prevent device destruction.

Heat produces two effects in solid-state lasers that are detrimental to laser action. The first effect, thermal population of low-lying LLLs resulting in re-absorption loss, can occur due to heat from quantum defects in the medium itself. The second effect, which can limit the ultimate power of a high-power laser, is thermal lensing of the amplifier medium itself.[3] Thermal lensing is an effect by which a temperature gradient within a solid-state laser material causes it to optically distort and so act as a lens. In the case of a solid-state laser, several effects are possible depending on the

A 2010 study by NIST researchers (technical note 1668) found that over 75% of green laser pointers emit dangerous levels of residual IR radiation. Many inexpensive devices lack an IR filter.

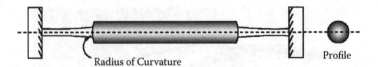

FIGURE 5.23   Thermal lensing in a solid-state laser rod.

geometry of the laser. This effect is most prominent in large solid-state lasers that are side-pumped and in which the rod is cooled from the outside surface. Heat, a result of quantum defect, builds from the center of the rod—the intensity is usually higher at the center of the rod, regardless, since the mode has highest intensity there—and is cooled from the outside of the rod, so a thermal gradient develops between the hotter central and cooler peripheral regions of the amplifier, resulting in both thermal stresses and a change of refractive index of the amplifier material usually resulting in the formation of a positive lens (i.e. as if the flat ends of the amplifier were actually convex). Thus changes to the optical cavity occur, as outlined in Figure 5.23, which ultimately reduce the output power.

Use of an amplifier medium with a small quantum defect can minimize the effects of thermal lensing, hence the interest generated around use of these materials. However, as outlined in this chapter, the same materials have other issues including re-absorption loss.

One approach to avoid thermal lensing and allow the scaling of solid-state lasers to high powers is the thin-disk laser,[4] in which the amplifier is a thin disk of solid-state material (usually of a few tenths of a millimeter thick), with an integral cavity mirror (at least on one side) and bonded directly onto a heatsink. Such lasers allow enormously powerful optical pump lasers to be used—so long as the heatsink is cooled there is essentially no limit—which, practically, can be used to produce small lasers with output powers of well over a kilowatt! The ability to cool the amplifier thoroughly across the gain medium minimizes thermal gradients.

A simple thin-disk laser is illustrated in Figure 5.24. Many practical thin-disk lasers use a more complex arrangement for pumping the gain medium, some consisting of a parabolic reflector around the OC with a hole in the middle for the optical

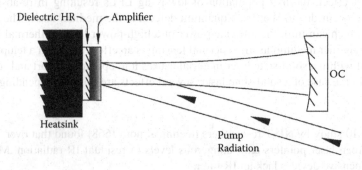

FIGURE 5.24   Thin-disk laser.

axis of the laser. In this case, unused pump radiation is returned to the amplifier again, increasing efficiency (since, not surprisingly, such a thin amplifier will not absorb all of the incident pump radiation).

Practically, thin-disk lasers have been produced using both vanadate ($Nd:YVO_4$) and Yb:YAG. The later material, pumped at 940nm, has a very small quantum defect—it oscillates at 1030nm—resulting in high efficiency but as a quasi-three-level material, cooling is important, which is why the thin-disk approach lends itself so well to use of this material.

REFERENCES

1. Fan, T.Y. and R.L. Byer. 1987. "Modeling and CW Operation of a Quasi-Three-Level 946nm Nd:YAG Laser." *IEEE Journal of Quantum Electronics*, QE-23, No 5.
2. Risk, W.A. 1988. "Modeling of Longitudinally Pumped Solid State Lasers Exhibiting Reabsorption Losses." *Journal of the Optical Society of America B*, Vol. 5 No. 7, July 1988.
3. Koechner, W. 1970. "Thermal Lensing in a Nd:YAG Laser Rod." *Applied Optics*, Vol. 9, No. 11, Nov 1970.
4. Antagnini, A. et al. 2009. "Thin-Disk Yb:YAG Oscillator-Amplifier Laser, ASE, and Effective Yb:YAG Lifetime." *IEEE Journal of Quantum Electronics*, QE-45, No. 8, 2009.

6 Generating Massive Inversions through Q-Switching

Short pulses of high peak power are extremely useful for a number of applications: While a long pulse may induce heat that penetrates deeply into a material, short pulses tend to ablate or vaporize the surface of a material in a manner that does not damage the substrate underneath. For this reason, short pulses find many applications in material processing and in the medical field.

While some pulsed lasers have naturally short pulses, most lasers do not. Q-switching is a technique for developing short pulses in those lasers. The technique is far more involved than simply switching the pump source on and off repeatedly, though—it involves storing energy in the gain medium and releasing it in one powerful burst. The result is a pulse of high peak power, with many small Q-switched lasers commonly producing pulses in the hundreds of kilowatts.

Covered in this chapter are the fundamentals of this technique, the technology used for practical switches, and several models for use with such lasers including a rudimentary model allowing prediction of pulse power based on the rates at which inversion and photons build in the cavity.

In a Q-switched laser, timing of the switching operation is important and a numerical model is developed here that describes the growth of inversion as a function of time. This model is generic and can be adapted to almost any Q-switched laser, including those with arbitrary pumping pulses (such as flashlamp pumped lasers).

6.1 INVERSION BUILDUP

In a continuous wave (CW) four-level laser, the onset of pumping generates an inversion that builds continuously until the threshold inversion is reached, at which point it is "capped" at this maximum value. Further pumping beyond this minimum value will result in excess gain, which, as we have seen in Chapter 3, "burns down" to the threshold value with the excess gain contributing to laser output. Inversion, then, never exceeds a threshold value (or so it appears).

To produce a pulsed laser output, one can switch pump power to the amplifier in a scheme known as gain switching. This will certainly produce a pulse, but there are three undesirable characteristics of the output pulse: (a) there is a significant delay between application of power and laser emission due to the required buildup of inversion up to and beyond threshold level, (b) the resulting pulse will have a relatively slow rise time while intra-cavity power builds, and (c) the maximum pulse power will never exceed CW values. In contrast to this, Q-switching produces pulses that are emitted almost immediately on command with a very fast rise time and, most importantly, with enormous peak pulse powers often exceeding hundreds of kilowatts.

Of course, gain switching does produce a pulse, and, in fact, oscillations in the output of a laser. It was discovered during development of the first ruby laser that the output consisted not of a single, smooth pulse but rather a series of "spikes"—a damped oscillation of the output that eventually reaches an equilibrium value. All solid-state lasers exhibit such behavior at turn-on, although the situation is more notable where the cavity decay time is short, certainly shorter than the time to build a substantial inversion. As such, most low-gain lasers do not exhibit such behavior, but high-gain solid-state lasers (including the YAG) do.

When pumping ensues, inversion (and hence gain) builds in the amplifier until inversion reaches the threshold value at which point one expects growth of the inversion to stop. This condition, which is the basis for many relationships developed so far in this text, is a steady-state condition. When the transient situation is examined, considering the first few microseconds after the pumping pulse starts, one finds that, at least momentarily, inversion can exceed the threshold value.

As depicted in Figure 6.1, in which a single pulse (in a series of such pulses produced) is examined in the transient view, although inversion (and hence gain) will have reached the threshold value, a negligible photon flux will have been built up in the cavity and so saturation will not immediately occur. Since pumping continues, inversion continues to build until eventually a photon flux powerful enough to reduce gain finally exists. By this time, though, the inversion will have reached a large value well in excess of the threshold value (assuming strong pumping). The large inversion thus serves to create a large output pulse. As the output pulse develops, the large

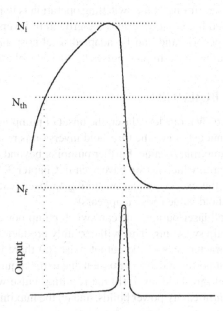

FIGURE 6.1 Spiking behavior of a gain-switched laser.

photon flux in the cavity decays more slowly than the inversion decays and so the inversion is depleted to a value lower than the threshold value. (This will also be a factor when developing an expression for pulse energy later in this chapter.) Since pumping is still applied, the cycle continues with the laser output oscillating in a series of repeating pulses. Pulse amplitude decreases as residual inversion increases with the laser eventually reaching steady-state output and the output stabilizing at steady-state values.

This "self-spiking" behavior is generally not desirable, but most solid-state lasers will exhibit this when pumping is first applied—it is often unnoticed since the laser reaches equilibrium in a short time (often under 100μs). This behavior may also be seen in the output of a laser that is perturbed, for example, through sudden cavity misalignment or a change in the pump power. A more useful, and controllable, method of producing pulses is Q-switching.

The idea behind Q-switching is simple: Lower the quality factor (or Q) of the cavity so that the laser cannot oscillate, then pump the amplifier allowing inversion to build. Since the laser cannot oscillate, inversion builds well beyond the aforementioned threshold value. When an output pulse is required, the Q of the cavity is restored, photon flux builds rapidly, and lasing ensues owing to the massive inversion that exceeds the threshold value, producing a massive pulse. A key requirement, then, is the Q-switch itself, which allows the quality of the cavity to be changed instantly to inhibit or allow lasing.

In theory, any laser can be Q-switched, but in practice to make it "worthwhile" the medium must have the ability to store a large amount of energy in the form of a large population at the ULL. To accomplish this, an ideal medium must have (a) a high density and (b) a long ULL lifetime. High density is required to allow the storage of significant ULL population before the Q-switch opens. Most gas lasers, for example a HeNe or argon-ion, have a low density which prohibits production of a sizeable output pulse. (In other words, gas lasers have a considerably lower extractable energy per unit volume than solid-state lasers.) In addition, a long ULL lifetime is required to allow the ULL population to build before natural decay serves to reduce it to an equilibrium level. (This is examined later in this chapter.)

In practice, then, the most practical lasers to Q-switch are solid-state lasers. Both three- and four-level lasers can be Q-switched. In the case of ruby, a three-level laser, a Q-switch can be used to allow a massive inversion to build following a single flashlamp pulse, producing a single massive pulse (although the inversion could be used to produce several pulses, each closely spaced in time). In the case of YAG, a four-level laser, both flashlamp- and continuously pumped lasers can be Q-switched; in the case of continuous pumping (such as a DPSS), the laser may well be able to produce pulses at a repetition rate of 10s of kHz.

Q-switching is thus called because the quality factor (Q) of the cavity is changed rapidly from a low-Q state where lasing is inhibited to a high-Q state where lasing begins.

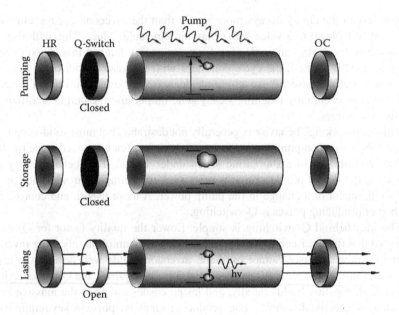

FIGURE 6.2 Sequence for Q-switching.

The basic sequence for production of a Q-switched laser pulse is outlined diagrammatically in Figure 6.2, with corresponding parameters such as inversion during these phases outlined in Figure 6.3. With the Q-switch closed, and therefore inhibiting lasing, the laser amplifier is pumped and a large inversion—one that is larger than the threshold inversion—builds. This inversion can be stored for a significant time—although certainly less than the lifetime of the ULL, unless continually

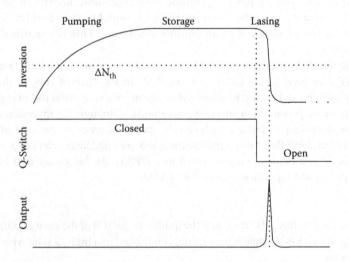

FIGURE 6.3 Timing for Q-switching.

pumped—and when an output pulse is required the Q-switch is opened allowing lasing to begin, at which point the massive inversion is rapidly depleted.

6.2   Q-SWITCH LOSS

No Q-switch is perfect—all practical switches have some amount of "leakage" and pass a portion of incident radiation even in the "fully closed" state. In order to be useful, a Q-switch must be able to prevent lasing even when the amplifier is generating maximum gain. When selecting a Q-switch, the first calculation completed is usually a determination of the minimum loss that the switch must induce into the laser cavity to prevent lasing. This will often determine the type of switch required.

Assuming the gain of the medium and the optical parameters of the laser are known, determination of the minimum loss proceeds in the same manner as previously described for the calculation of mirror reflectivities. A gain threshold equation is developed for the specific laser incorporating the loss of the Q-switch as an unknown quantity. The threshold gain is then equated to the small-signal gain of the medium and the equation solved for the switch loss. This will be the maximum allowed transmission of the Q-switch to prevent lasing (i.e. the required "holdoff" of the Q-switch).

EXAMPLE 6.1   MINIMUM LOSS OF A Q-SWITCH

To illustrate the loss a Q-switch must induce, we consider two lasers: a ruby and a YAG.

In the case of the ruby, the small-signal gain is often quoted as $20m^{-1}$. This is a reasonable figure for a ruby, but as an exercise, let's calculate the absolute maximum possible small-signal gain.

Ruby consists of 0.05% Cr^{3+} ions in a host of Al_2O_3. The doping density is hence $1.58*10^{25}m^{-3}$. Assuming a 100% inversion (that would, after all, yield maximum gain), the maximum gain can be calculated according to Equation (2.20) as:

$$g_{MAX} = \sigma_0 \Delta N = 2.5 \times 10^{-24} m^2 \times 1.58 \times 10^{25} m^{-3} = 39.5 m^{-1}$$

This is, of course, an absolute maximum, and real lasers will never be able to reach this level of inversion—the largest ruby lasers can achieve a maximum gain of about $25m^{-1}$.

Now assume a ruby laser with the following parameters:

$R_{HR} = 100\%$
$R_{OC} = 90.0\%$
$x = 7.5cm$
$\gamma = 0.1m^{-1}$

The maximum allowed transmission of the Q-switch can then be computed in the same manner as previously done in Example 2.7 for optic reflectivity (i.e. by equating g_{th} to the maximum gain of the medium and solving for the maximum transmission of the switch):

$$g_{th} = 25m^{-1} = \gamma + \frac{1}{2x} \ln\left(\frac{1}{R_1 R_2 T_Q^2}\right)$$

The Q-switch must transmit no more that 16.3% in the "closed" state to completely inhibit lasing and hence operate properly.

Now consider a YAG laser that has a smaller gain of $10m^{-1}$. Being a four-level laser, the approach of using 100% inversion is invalid. Only a small percentage of the total number of lasing atoms (or ions, in this case) are ever involved in the lasing process.

Consider the following parameters:

R_{HR} = 100%
R_{OC} = 90.0%
 x = 10cm
 γ = 0.1m^{-1}

Using the same approach as above, the maximum transmission of the Q-switch is now calculated to be 39%.

Armed with this minimum inserted loss figure for each laser, we may now proceed to choose a suitable switch.

6.3 AOM SWITCHES

Acousto-optic modulators (AOMs) work by diffracting an incoming beam, which in turn induces a loss when the device is inserted into a laser cavity, as demonstrated in Figure 6.4, which shows an actual AOM device in a laser cavity. A piezoelectric crystal driven by a radio-frequency (RF) source produces an acoustic wave in a crystal substrate. As the wave travels through the material (at the acoustic velocity, which is a property of the crystal), it produces compressions and rarefactions in the crystal, which subsequently induces localized changes in the refractive index: where the crystal is compressed, the index increases, and where rarefaction occurs, the index decreases. This in effect creates "bands" of high and low indices inside the crystal, creating a grating that diffracts incident light. The end of the device is cut at an odd angle to discourage formation of standing waves in the device. An AOM is diagrammed in Figure 6.5—the

Acousto-optic (AO) Q-switches are by far the most common active switch, given the huge cost advantage over electro-optic (EO) switches.

FIGURE 6.4   An AOM with the cover removed.

top diagram shows diffraction resulting from light passing through the device and the lower diagram shows placement in the cavity at the Bragg angle (explained below).

While the resulting diffraction can resemble that of a "traditional" ruled transmission grating (most notably with diffraction occurring on both sides of the zeroth order beam in what is called *Raman-Nath* diffraction), the highest efficiency (that is, induction of the largest loss into the zeroth-order beam) occurs when *Bragg*

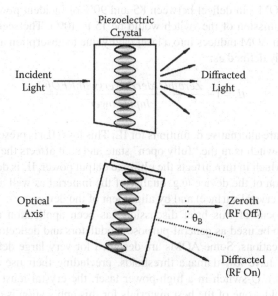

FIGURE 6.5   Diffraction in an AOM and placement in a laser cavity.

diffraction occurs. To accomplish this, the AOM is placed inside the laser cavity at an angle of:

$$\theta_{BRAGG} = \frac{\lambda f}{2v} \tag{6.1}$$

where λ is the wavelength of the incident radiation, f is the RF frequency, and v the acoustic velocity of the material. This is the Bragg angle in radians: The formula shown is already simplified by using the small-angle theorem in which $\sin\theta = \theta$ and uses external angles as well. (The reader is cautioned that some texts and datasheets refer to internal angles inside the device that must be converted to external angles through application of Snell's law.)

The loss induced in Raman-Nath operation is much less than that induced when aligned for Bragg operation, with the exact type of operation required dictated by the loss required (as computed in Section 6.2). Losses of up to 90% are possible with an AOM, but Bragg mode must be used to accomplish this, and alignment of this angle is quite critical. (AOM mounts, such as that in Figure 6.4, feature a precision adjustment for this angle.)

The performance of AOM devices is characterized by two primary parameters: diffraction efficiency (DE) and insertion loss (IL). Diffraction efficiency quantizes how well an AOM can deflect power in the central beam (the zeroth order) to the first order and is defined as:

$$DE = \frac{FirstOrderPower(withRFOn)}{ZerothOrderPower(withRFOff)} \tag{6.2}$$

Typically, an AOM can deflect between 85 and 90% of incident power (and so, the minimum transmission of the switch would be 15 to 10%). The second parameter, IL, is the loss an AOM induces into a laser cavity due to absorption and other losses and is commonly defined as:

$$IL = 1 - \frac{ZerothOrderPower(withRFOff)}{InputPower} \tag{6.3}$$

Note that there are alternative definitions for IL. This loss (IL) is present in the cavity even when the switch is in the "fully open" state and so it affects the threshold gain of the system, which in turn affects the ultimate output power. IL is determined both by the fabrication of the device (e.g. quality of the material as well as antireflective coatings on the crystal surface) and by alignment of the device.

While the focus of this brief discussion has been application as a Q-switch, AOMs may also be used as general-purpose modulators and deflectors for a host of photonics applications. Some AOMs are designed for very large deflection angles, but often these have low damage thresholds, precluding their use as Q-switches. To be useful as a Q-switch in a high-power laser, the crystal must feature a high damage threshold—one of the best materials for this application is quartz or fused silica—and crystal surfaces must be coated with antireflective coatings, also with a high damage threshold.

EXAMPLE 6.2   BRAGG ANGLE IN A Q-SWITCH

The Q-switch for a typical high-power YAG laser is manufactured from fused silica, which features an extraordinarily high damage threshold as required for intra-cavity usage. These devices require medium-power RF signals (usually between 10W and 100W) to operate.

The parameters for a "typical" fused silica Q-switch are f = 27MHz and v = 3760m/s.

Assuming it is used inside the cavity of a YAG laser (1064nm), the switch must be inserted into the laser cavity at an angle predicted by Equation (6.1):

$$\theta_{BRAGG} = \frac{\lambda f}{2v} = \frac{1064 \times 10^{-9}\, m \times 27 \times 10^6\, Hz}{2(3760 m/s)} = 0.00382 radians$$

In other words, at the Bragg angle for maximum diffraction efficiency. Q-switches usually feature adjustments allowing precise selection of angle.

EXAMPLE 6.3   AOM DEFLECTION

Tellurium oxide (TeO_2) AOMs are small and operate with low RF drive power (often 1W or less). These devices find application as modulators and deflectors. When used as a deflector, the separation angle (between the beam in the "RF off" state and the "RF on" state) is $2\theta_{BRAGG}$ as shown in Figure 6.5.

Consider a small TeO_2 AOM with an acoustic velocity of 650m/s used to deflect a HeNe beam at 632.8nm. Assuming the RF drive signal can be varied between 50MHz and 100MHz, the range of angular deflection possible ranges between:

$$\theta_{SEPARATION} = 2\theta_{BRAGG} = \frac{\lambda f}{v} = \frac{632.8 \times 10^{-9}\, m \times 50 \times 10^6\, Hz}{650 m/s} = 0.0487 radians$$

and

$$\theta_{SEPARATION} = 2\theta_{BRAGG} = \frac{\lambda f}{v} = \frac{632.8 \times 10^{-9}\, m \times 100 \times 10^6\, Hz}{650 m/s} = 0.0974 radians$$

with the resulting range of deflection being just under three angular degrees.

FIGURE 6.6   An electro-optic modulator with analyzer (foreground) in a test fixture.

6.4   EOM SWITCHES

EOM switches are based on an EO effect in which the index of refraction of a material changes in the presence of a strong external electric field. Polarized radiation (either polarized explicitly, for example, by an intra-cavity Brewster window or polarizing beamsplitter, or perhaps by the medium itself since some solid-state media such as ruby and alexandrite are naturally polarized) passes through an EO crystal called a *Pockels cell* and is incident on a second, crossed, polarizer (called the *analyzer*). An EOM of this type, on a test jig, is shown in Figure 6.6. The actual EOM is the small cylinder on the right: it is a relatively small crystal held within a cylinder that acts as electrical insulation from the high voltages required. (The cylinder is often filled with insulating fluid.) Also visible in this figure on the left (in the foreground) is a stack of Brewster plates used as an analyzer. The analyzer consists of six plates, of which three are visible in the photo. (They are normally covered by the metal tube and so not visible.) More plates may be added to effect a larger intra-cavity loss in one polarization and decrease transmission of this switch in the "closed" state.

The device is set up as per Figure 6.7 such that the analyzer and polarizer are crossed and so in the unenergized state, no change in the polarization occurs in the crystal and light does not pass through the device—the switch is essentially "closed." When energized, the cell "twists" the polarization of radiation passing through it— with the optimal applied voltage, a "twist" of a half-wavelength (or 90 degrees) occurs and the device passes radiation—this is the "open" state. In this manner, the Pockels

Use of EO switches is limited in the infrared since use of longer wavelengths requires higher voltages that must not exceed the rating of the device.

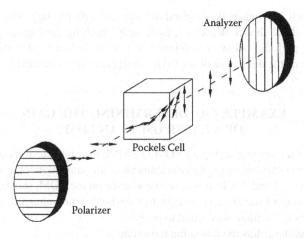

FIGURE 6.7   Polarization of radiation in an EOM.

cell may be pictured as a waveplate that is voltage controlled: the phase-delay of radiation passing through the device can be varied according to the applied voltage. This also means that the EOM is a variable modulator in which the transmission of the device can be varied according to the applied voltage (with zero volts resulting in zero transmission).

An important parameter of a Pockels cell is the half-wave voltage (denoted $V_{1/2}$), the voltage at which a polarization "twist" of exactly 90 degrees occurs and so maximum transmission from a switch configured with crossed polarizers (as per Figure 6.7) results. This voltage can range from hundreds to thousands of volts for many devices.

Half-wave voltage is a function of wavelength, and longer wavelengths (i.e. IR) require larger half-wave voltages than shorter wavelengths—this is often the limiting factor when using an EOM device with long IR wavelengths. The half-wave voltage for EOM devices is usually rated at one specific wavelength, but not always the wavelength of interest, and so a simple conversion must be applied as follows:

$$\frac{V_{\frac{1}{2}-1}}{V_{\frac{1}{2}-2}} = \frac{\lambda_1}{\lambda_2} \tag{6.4}$$

The transmission of the switch is computed according to:

$$T = T_0 \sin^2\left(\frac{\pi}{2} \frac{V}{V_{\frac{1}{2}}}\right) \tag{6.5}$$

where T_0 is the maximum transmission of the Q-switch in the "fully opened" state (i.e. factoring in insertion loss), V is the applied voltage, and $V_{1/2}$ the half-wave voltage at the wavelength employed.

Note as well that when the applied voltage exceeds the half-wave voltage the transmission of the device decreases, eventually reaching zero again at twice that voltage. This periodic behavior continues as voltage increases, but the maximum voltage rating of the device, or of the driver, will serve as a practical limit.

EXAMPLE 6.4   DETERMINING THE GAIN OF A LASER USING AN EOM

The gain of a laser employing an EOM Q-switch may be determined by using the EOM as a variable intra-cavity attenuator in the same manner as described in Examples 2.4 and 2.5. In this case, the voltage on the EOM, starting at zero, is increased until the laser is brought to threshold—the voltage at which the laser begins to oscillate was found to be 9kV.

The YAG laser has the following parameters:

$R_{HR} = 100\%$

$R_{OC} = 90.0\%$

$x = 15cm$

$\gamma = 0.1m^{-1}$

The Pockels cell has the following parameters: the insertion loss is 5.6% and the half-wave voltage is 7.62kV at 632.8nm.

First, the half-wave voltage is found at the wavelength of interest using Equation (6.4):

$$\frac{V_{\frac{1}{2}-1}}{7.62kV} = \frac{1064nm}{632.8nm}$$

The half-wave voltage at the lasing wavelength of 1064nm is hence found to be 12.8kV. A high voltage such as this is a reasonable expectation since IR wavelengths require a higher voltage than visible wavelengths to "twist" the polarization by the same amount.

Next, the transmission of the switch is calculated at threshold using Equation (6.5), in which the maximum transmission of the switch in the "fully opened" state is $1 - 0.056 = 0.944$:

$$T = 0.944 \sin^2\left(\frac{\pi}{2}\frac{9.0}{12.8}\right) = 0.753$$

Finally, a gain threshold equation is formulated for the laser and the threshold gain, now the small-signal gain, is calculated.

$$g_0 = \gamma + \frac{1}{2x}\ln\left(\frac{1}{R_{HR}R_{OC}T_Q^2}\right) = 0.1 + \frac{1}{2(0.15m)}\ln\left(\frac{1}{0.9(0.753^2)}\right) = 2.34m^{-1}$$

Where an EOM is not perfectly aligned (i.e. the analyzer and polarizer are not perfectly perpendicular to each other), the "apparent" half-wave voltage can be significantly different from that expected. An angular offset results in a small modification to the transmission equation (Equation 6.5) as follows:

$$T = T_0 \sin^2\left(\frac{\pi}{2}\left(\frac{V}{V_{\frac{1}{2}}} + \phi\right)\right)$$

(6.6)

Of course, a misalignment such as this results in a "leakage" transmission even when the device is in the closed/unenergized state, which can be computed by substituting a zero voltage in the above equation. Depending on the gain of the laser involved, this may affect the suitability of the EOM for a particular system since the EOM may not induce enough loss in this misaligned state to completely prevent oscillation.

EXAMPLE 6.5   AN IMPERFECTLY ALIGNED EOM

A Pockels cell EOM with a half-wave voltage of 8.35kV (already compensated for the actual wavelength of the laser of 694nm by Equation 6.4) was aligned crudely in a laser. Test-firing at various voltages revealed that maximum output was observed when a voltage of 5.7kV was applied to the device. The offset of the polarizer and analyzer can hence be computed using Equation (6.6) as follows:

The maximum output occurs when:

$$\frac{\pi}{2}\left(\frac{V}{V_{\frac{1}{2}}} + \varphi\right) = \frac{\pi}{2}$$

Substituting for V and $V_{1/2}$:

$$\frac{\pi}{2}\left(\frac{5.7kV}{8.35kV} + \varphi\right) = \frac{\pi}{2}$$

So $\phi = 0.3174$ and the offset is ($\phi\pi/2 =$) 0.4985 radians or 28.6 degrees, meaning the polarizer and analyzer are crossed at an angle of 61.4 degrees, not 90 degrees as expected. When considering the transmission of the switch as a function of applied voltage, the following formula must be used for this specific Q-switch:

$$T = T_0 \sin^2\left(\frac{\pi}{2}\left(\frac{V}{8350} + 0.3174\right)\right)$$

where V is the applied voltage. Of course, misalignment in this manner also means the transmission in the "closed" state (with zero volts applied to the switch) will in this case be 21.6%—better alignment of the switch is required for higher induced losses.

In many cases, half-wave voltage is not quoted on the datasheet for an EOM but rather quarter-wave voltage is, where this is the voltage to effect a 45-degree "twist" in light passing through the device. Such configurations are used to reduce the required voltage since the polarizer and analyzer are crossed at 45-degree, not 90-degree, angles.

Use of an EOM as a Q-switch requires a pulse generator with a fast rise-time. In the past, the standard method was to use a high-voltage vacuum-tube switch, such as a thyratron, that could apply a voltage to the switch rising from zero volts to the half-wave voltage in a few nanoseconds (or less). Today, fast solid-state electronic switches are more common, though the voltage ratings of these devices are lower than their vacuum-tube counterparts, which is why some are used in quarter-wave mode—although there is a trick to this arrangement.

If the same arrangement were employed as shown in Figure 6.7, i.e. with one polarizer and one analyzer arranged at a crossing-angle of 45 degrees, and the Pockels cell in the middle, the maximum possible loss that would result when the quarter-wave voltage is employed is 50% in the "closed" state—a transmission of 50% might not be enough to Q-switch many high-gain lasers that require higher inserted losses. Today, use of quarter-wave mode in many high-gain solid-state lasers is common, however, with this application made possible through the addition of a quarter-wave plate between the Pockels cell and one cavity mirror. As seen in Figure 6.8, polarized intra-cavity radiation (which may require the addition of a polarizer for non-polarizing media such as YAG) passes through a Pockels cell (assuming zero voltage is applied to the cell), resulting in no polarization change. This radiation then passes through a quarter-wave plate, reflects from the HR, and passes through the quarter-wave plate again resulting in a half-wave (or 90-degree) polarization change. This polarization is opposite the original polarization of the amplifying medium (or the intra-cavity polarizer, if so used), so losses will be large and lasing will be inhibited. With a quarter-wave voltage applied to the cell, radiation is twisted by a quarter-wave upon each pass through the cell (with radiation making two passes through the cell

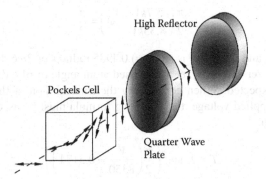

High Reflector

Pockels Cell

Quarter Wave
Plate

FIGURE 6.8   An EOM in quarter-wave mode.

EOMs are characterized by the half- or quarter-wave voltage of the device with quarter-wave mode often used to reduce the operating voltage.

in total) as well as a polarization change upon each pass through the quarter-wave plate resulting in an overall polarization change of one wave, hence no net change in polarization has occurred and so the laser will oscillate.

Alternative arrangements are also possible where the laser fires when the Q-switch has zero volts applied. By eliminating the quarter-wave plate in Figure 6.8, polarized intra-cavity radiation will be rotated 45 degrees upon passing through the cell and another 45 degrees on the return path for a total change of 90 degrees. Lasing will thus be inhibited when the quarter-wave voltage is applied to the cell, and the laser will oscillate when zero volts are applied (so that no polarization change occurs on transit through the cell). This arrangement is called *off Q-switching* or *pulse off Q-switching*.

6.5 PASSIVE Q-SWITCHES

Both AO and EO switches are active, meaning they open on application of an electrical signal, allowing precise timing of the emission of a Q-switched pulse. A simpler type of switch is a passive Q-switch that, when inserted into the cavity of a laser (as shown in Figure 6.9), will open when a threshold for intra-cavity radiation is reached, allowing emission of a laser pulse.

The most common form of passive switch is the *saturable absorber*, and physically this device can be a cell of liquid dye, a solid material such as a crystal, or a thin-film on a glass substrate. The saturable absorber is inserted into the laser cavity and as such increases the loss for the laser. As shown in Figure 6.10, the absorber is configured such that the laser will still oscillate although with an enormous threshold value that is far above the "normal" threshold without the switch inserted. As pumping ensues, inversion builds until the laser finally begins to oscillate (and a large inversion will have been built up to allow this), and intra-cavity radiation then builds in the cavity and quickly saturates or "bleaches" the absorber so that it, in effect, becomes transparent, lowering losses and allowing utilization of the large inversion that has built to produce a pulse. The absorber remains transparent until radiation in the cavity

FIGURE 6.9 A passive Q-switch in a YAG laser cavity.

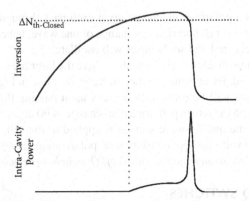

FIGURE 6.10   Inversion in a passively Q-switched laser.

decays, at which point the absorber returns to the original high-loss state and the cycle may be repeated. (The Q-switch is hence "closed" again and will remain so until the high threshold is once again exceeded to allow oscillation to begin.) In practice, the pump power of the laser amplifier (and hence gain) must be raised to allow the medium to develop a gain approximately 10% above the threshold of the cavity with the switch inserted in the unsaturated or closed state. This allows the laser to begin oscillation with the absorber in the cavity, which in turn is necessary to allow intra-cavity power to build and hence saturate the absorber.

EXAMPLE 6.6   A PASSIVE Q-SWITCH

A passive Q-switch is to be used with a YAG laser. The transmission of the Q-switch is measured at the same wavelength as the laser (1064nm) using a low-power (1 milliwatt) CW laser so as not to saturate the device and obtain a false reading. The transmission under such conditions was found to be 68%. The transmission of the same Q-switch was determined in the "open" state by repeating the same measurement using the output pulse from a Q-switched laser and was found to be 90%. The Q-switch, then, exhibits a 32% inserted loss in the "closed" state and a 10% loss in the "open" state. (A portion of the loss in the "open" state is likely due to reflections from the surfaces of the device that are not antireflective-coated at the 1064nm wavelength used, as well as absorption from the excited state in the absorber.)

The laser proposed for use with this switch is a diode-pumped YAG laser with the following parameters:

$R_{HR} = 100\%$
$R_{OC} = 90.0\%$
$x = 108$mm
$\gamma = 0.1$m^{-1}

The gain of this laser was determined experimentally by inserting a variable attenuator wheel into the cavity in the same manner as described in Example 2.5 and was found to be 4.58m^{-1}.

With the passive switch inserted into the laser cavity, the laser must still be able oscillate since intra-cavity radiation is required to saturate the absorber and allow pulse output. The minimum transmission of the inserted switch is found using the same methodology as outlined in Chapter 2:

$$T = \left(\sqrt{R_{HR} R_{OC}} e^{(g-\gamma)2x} \right)^{-1}$$

where g is 4.58m^{-1}. The resulting minimum transmission of an inserted optical element to allow oscillation is hence 65%. Since the transmission of the Q-switch (determined experimentally to be 68%) is only slightly larger than this value, the laser will barely begin to oscillate when the laser medium approaches maximum inversion but not enough to passively Q-switch this laser properly, for which the gain with the Q-switch inserted must be about 10% higher, which allows a large intra-cavity power to develop and saturate the device. This was verified experimentally in which a very low level output was seen instead of the expected massive pulse.

Now consider using the same Q-switch with a different solid-state laser: a flashlamp-pumped unit with much higher gain. Using the same method outlined in Example 2.5 with that laser, the gain of the amplifier was determined to be 11.5m^{-1}.

Use the following parameters:

R_{HR} = 100%
R_{OC} = 90.0%
x = 5cm
γ = 0.1m^{-1}

The minimum transmission of the inserted switch is found using the same methodology to be 59.6%. Since the transmission of the Q-switch (determined experimentally to be 68%) is considerably larger than this value, the laser will begin to oscillate when the laser medium approaches maximum inversion (and hence gain), at which point the switch will saturate and open, producing a Q-switched output pulse.

With this Q-switch installed, gain in the laser amplifier will build to 8.87m^{-1} (again, determined using the equation developed in Example 6.1, in which the "closed" switch transmission of 0.68 was used for T_Q), at which point the switch will saturate and open, with the threshold gain of the system falling to 3.26m^{-1}. (A T_Q value of 0.90 is used since the "open" switch has a transmission determined to be 90% under these high-intensity conditions, due both to residual absorption of the now saturated switch and to reflections from the surfaces of the switch material itself that were not coated for the exact same wavelength employed.) The emission of a massive Q-switched pulse is observed with this configuration. These two gain figures ("initial" and "open") may then be converted to values of inversion and used with the relationship developed in the next section to determine actual pulse power and energy.

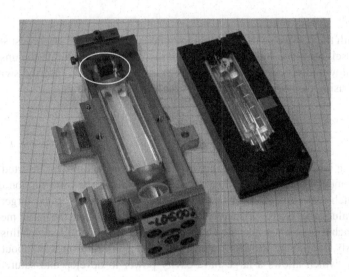

FIGURE 6.11   A small passively Q-switched laser with flashlamp.

Although shown in the cavity of a large laser in Figure 6.9, passive Q-switches are more commonly found on small solid-state lasers for medical and instrumentation use where small size (or simplicity) is a huge benefit. Lasers used for LIBS (Laser Induced Breakdown Spectroscopy) are frequently compact units employing solid-state passive Q-switches. Other applications where passive Q-switches are frequently employed are small medical lasers used for eye surgery (for example, photocoagulation and photodisruption). Passive Q-switches are simple and compact, allowing construction of small solid-state lasers with the desirable qualities of a Q-switched output (namely a short pulse duration that allows vaporization of a precise amount of tissue) to be built—a compact laser is all that is required since many of these procedures require a pulse energy of under 10mJ with a pulse length under ten nanoseconds. The compact size allows simplification of the optical path (and hence optical elements and associated alignment requirements) within the product.

A small laser of this type is shown in Figure 6.11, in which a small YAG rod 50mm in length is pumped by a small flashlamp (shown on the right). The actual passive switch is at the top of the photo and is circled—it is a small solid-state crystal placed within the cavity, the mirrors of which are shown in mounts on the end of the laser.

While a variety of media can be used (with a cell of saturable dye being the most popular in past designs), a common material for a solid-state switch is chromium-doped YAG (Cr^{4+}:YAG). This material, which is visibly dark orange in color, is usable in the near-IR region between 800nm and 1200nm. It can be designed for any transmission required, and the contrast (the ratio of absorption in the initial "closed" state to the absorption in the saturated "open" state) is approximately 10 so that a switch that absorbs 50% of radiation initially will, when saturated, absorb only 5%. Contrast is one of the most important parameters when considering a passive

Q-switch, and it is similar to DE for an AOM in characterizing the efficiency of the switch, and will determine the gain (and inversion) for a passively Q-switched laser in both the initial and saturated states, which will ultimately determine the output power (as covered later in this chapter).

Aside from Cr^{4+}:YAG, numerous other saturable materials exist for passive Q-switching including Gadolinium-scandium-gallium garnet (GSGG) doped with chromium and magnesium (GSGG:Mg:Cr^{4+}), tri-valent vanadium doped YAG (V^{3+}:YAG), Cobalt (Co^{2+}) doped $MgAl_2O_4$ (Co:MALO or Co:spinel), and a host of other materials. All work is in the near-IR region, where the vast majority of solid-state lasers operate.

Compared to active switching (using an AOM or EOM), passive switching does not allow triggering of an output pulse—it simply appears when the inversion has built to a large enough level, which is dependent on the rate of pumping—and the losses of the switch result in lower power output. Still, the simplicity of the method, which does not require driver electronics of any kind, lends itself well to the design of simple, compact, and reliable Q-switched lasers.

6.6 A MODEL FOR PULSE POWER

Regardless of the technology of the Q-switch employed, one of the most useful parameters to be able to predict is the power and energy of the resulting pulse. In most cases, the initial gain (and hence initial inversion) and the threshold gain (and hence threshold inversion) are known, so a model will be developed based on these known parameters. In a simple view of Q-switching, the pulse begins with a large initial inversion, the switch opens, initiating the laser pulse, and inversion quickly decays as the pulse is produced, with the inversion (and the pulse) terminating when threshold is reached. This is a naïve view since, as discussed in Section 6.1, a large intra-cavity photon flux will have been built up and will serve to reduce the population even further than expected—the final inversion will fall to a level considerably lower than the threshold value. The relationship between inversion and the resulting pulse energy is anything but straightforward, and is outlined in Figure 6.12, in which the inversion starts at a very large initial value (N_i) and with essentially zero intra-cavity power. Cavity power builds until the rate of stimulated emission becomes high. In turn, the large intra-cavity flux quickly depletes the inversion, reaching a peak when the threshold inversion is reached. Because a large intra-cavity power exists, the inversion is further depleted reaching the final value (N_f) as the intra-cavity power decays and the pulse terminates.

Development of a rudimentary relationship between the initial and final inversions and the energy of a Q-switched pulse requires that both inversion and the flux of photons produced be considered together, and we begin by formulating two

As well as the common Q-switches outlined here, mechanical switching in the form of a rotating mirror was used in early designs.

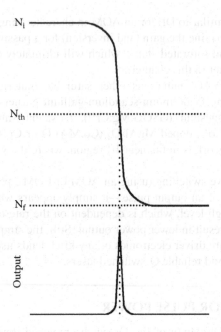

FIGURE 6.12 Inversion and resulting pulse power as a function of time.

equations that describe both the rate at which photons build during the Q-switched pulse and the rate at which inversion changes during that same pulse. In describing the rate at which photons build, we introduce the term n, which represents the number of photons inside the laser (per unit volume) at any time t. (So far we have considered photon flux ρ, which is the number of photons per second per unit area; n is simply the number of photons inside the cavity per unit volume.)

From the Rigrod model in Chapter 4, intensity of a stream of photons in a laser cavity is seen to build according to:

$$\frac{dI_+}{dz} = g(z)I_+ \tag{6.7}$$

This stream of photons traverses the amplifier at a speed of:

$$v = \frac{dz}{dt} = \frac{c}{n_{INDEX}} \tag{6.8}$$

which is the speed of light inside the medium of the amplifier (where n_{INDEX} in the above equation is actually the index of refraction of the amplifier material, not number of photons as it is for the rest of this development). The intensity of photons inside the amplifier is then seen to grow over time at the rate of:

$$\frac{dI}{dt} = gvI \tag{6.9}$$

where the direction I_+ was dropped. Consider now that the rate at which the number of photons grows (dn/dt) is the same as the rate at which intensity grows, so the input to dn/dt will be substituted for dI/dt. For a complete rate equation, then, one must consider the rate of generation of photons (as above) and the rate of decay of photons in the cavity to yield:

$$\frac{dn}{dt} = IN - OUT = gvn - \frac{n}{\tau_{CAVITY}} \tag{6.10}$$

where the rate of cavity loss is expressed as a function of the *photon lifetime* in the cavity, as previously developed in Section 3.11. Furthermore, we may eliminate v by using Equation (3.50) from Section 3.11:

$$\tau_{CAVITY} = \frac{1}{c \times g_{th}} \tag{6.11}$$

where c is actually the velocity in the amplifier medium (i.e. v in our equations here). Equation (6.10) thus simplifies to:

$$\frac{dn}{dt} = \frac{g}{g_{th}\tau_{CAVITY}} n - \frac{n}{\tau_{CAVITY}} \tag{6.12}$$

and furthermore to:

$$\frac{dn}{dt} = \frac{n}{\tau_{CAVITY}} \left(\frac{g}{g_{th}} - 1 \right) \tag{6.13}$$

Knowing that gain is directly proportional to inversion (from Equation 2.20), this same rate can be expressed in terms of inversion as:

$$\frac{dn}{dt} = \frac{n}{\tau_{CAVITY}} \left(\frac{\Delta N}{\Delta N_{th}} - 1 \right) \tag{6.14}$$

This expresses the rate of change of photons inside the cavity, where n in the above equation is the number of photons. In reality it is a function of time $n(t)$, just as inversion is $\Delta N(t)$ as well.

The second equation, describing the rate at which inversion changes during the Q-switched pulse, can be formulated by first knowing that a single stimulated emission event (i.e. the production of a lasing photon) results in a decrease of two atoms in the inversion since the ULL decreases by one atom but the LLL increases by one atom simultaneously. (Since any practical laser system has a finite LLL lifetime that is longer than the average Q-switched pulse, it can be assumed that the atom reaching the LLL stays in that level for the duration of the pulse.) The rate of change of inversion, then, can be expressed as:

$$\frac{d\Delta N}{dt} = IN - OUT = 2\left(\frac{N_{PUMP}}{\tau_{PUMPLEVEL}} - gvn \right) \tag{6.15}$$

It may be logically assumed that decay from the pump level (which contributes to increasing the inversion) will be quite negligible during the short time frame of the Q-switched pulse, so this term is eliminated. The remaining term is expressed as a function of actual and threshold gain in the same manner that Equation (6.12) was simplified as:

$$\frac{d\Delta N}{dt} = -2\left(\frac{g}{g_{th}\tau_{CAVITY}}\right)n \qquad (6.16)$$

And once again this can be expressed in terms of inversion as:

$$\frac{d\Delta N}{dt} = -2\frac{n}{\tau_{CAVITY}}\left(\frac{\Delta N}{\Delta N_{th}}\right) \qquad (6.17)$$

The reader is again reminded that the number of photons in the cavity, n, and the inversion, ΔN, are both time-dependent and so should most correctly be written as $n(t)$ and $\Delta N(t)$, respectively.

Dividing Equation (6.14) by Equation (6.17), we may obtain a rate relationship between photons in the cavity and inversion as follows:

$$\frac{dn}{d\Delta N} = \frac{\dfrac{n}{\tau_{CAVITY}}\left(\dfrac{\Delta N}{\Delta N_{th}} - 1\right)}{-2\dfrac{n}{\tau_{CAVITY}}\left(\dfrac{\Delta N}{\Delta N_{th}}\right)} = -\frac{1}{2} + \frac{1}{2}\frac{\Delta N_{th}}{\Delta N} \qquad (6.18)$$

This equation is then integrated to solve for n as follows:

$$n = \frac{1}{2}\Delta N_{th}\ln(\Delta N) - \frac{1}{2}\Delta N + k \qquad (6.19)$$

To solve for the integration constant, the boundary condition is applied where the Q-switch is just opened. At this point, $\Delta N = \Delta N_i$ (i.e. the initial inversion) and the number of photons in the cavity (n) is zero so that:

$$\frac{1}{2}\Delta N_{th}\ln(\Delta N) - \frac{1}{2}\Delta N + k = 0 \qquad (6.20)$$

$$k = -\frac{1}{2}\Delta N_{th}\ln(\Delta N_i) + \frac{1}{2}\Delta N_i \qquad (6.21)$$

Thus, the complete solved expression for the number of photons in the cavity is:

$$n = \frac{1}{2}\Delta N_{th}\ln(\Delta N) - \frac{1}{2}\Delta N - \frac{1}{2}\Delta N_{th}\ln(\Delta N_i) + \frac{1}{2}\Delta N_i \qquad (6.22)$$

$$= \frac{1}{2}\Delta N_{th}\ln\left(\frac{\Delta N}{\Delta N_i}\right) - \frac{1}{2}\Delta N + \frac{1}{2}\Delta N_i$$

Knowing the number of photons in the cavity as well as cavity parameters yields a simple expression for the power of a Q-switched pulse:

$$P = (1 - R_{OC})h\upsilon \frac{nV}{\tau_{CAVITY}} \tag{6.23}$$

where $(1 - R_{OC})$ is the portion that passes through the OC, $h\upsilon$ is the energy of an individual photon, nV is the number of photons in the cavity (where V is the volume, since n is in "per unit volume" units), and the reciprocal of the photon lifetime in the cavity is the rate at which photons decay. Expanding this equation by substituting for n,

$$P = (1 - R_{OC})h\upsilon V \frac{1}{\tau_{CAVITY}} \left[\frac{1}{2} \Delta N_{th} \ln\left(\frac{\Delta N}{\Delta N_i} \right) - \frac{1}{2} \Delta N + \frac{1}{2} \Delta N_i \right] \tag{6.24}$$

This a time-dependent quantity (i.e. $P(t)$) since n is a time-dependent quantity (although it is not shown as such for clarity). If one realizes that the peak output of the laser occurs when inversion reaches the threshold level, the peak power can be expressed as:

$$P_{PEAK} = (1 - R_{OC})h\upsilon V \frac{1}{\tau_{CAVITY}} \left[\frac{1}{2} \Delta N_{th} \ln\left(\frac{\Delta N_{th}}{\Delta N_i} \right) - \frac{1}{2} \Delta N_{th} + \frac{1}{2} \Delta N_i \right] \tag{6.25}$$

Finding the energy should be a simple matter of integration of the power equation (Equation 6.25) between the limits of N_i (the initial inversion) and N_f (the final inversion). In reality, while ΔN_i is often easily found from measurements of maximum gain, the final inversion often requires an approximation. In this case, where $\Delta N_i \gg \Delta N_{th}$, the final term ($\frac{1}{2} \Delta N_i$) will dominate and so by inspection it can be seen that this is the maximum number of photons in the cavity as well. The resulting energy of the pulse is hence approximated as the number of photons (where each inversion generates a photon) multiplied by the energy per photon and the fraction emitted through the OC as:

$$E = \frac{1}{2}(1 - R_{OC})h\upsilon\Delta N_i \tag{6.26}$$

where ΔN_i is in "absolute units" and so, if provided in "per unit volume" units, must first be multiplied by volume.

6.7 MULTIPLE PULSE OUTPUT

Normally, Q-switching involves production of a single pulse using essentially all available inversion, but it is possible to produce multiple pulses from the same stored inversion in a multiple pulse scheme employing an EOM switch. A number of analytical techniques including holography and particle imaging velocimetry require production of two or more identical pulses spaced at a fixed, but short, time interval apart. The basic idea is to partially open the Q-switch for the first pulse, depleting part but not all of the stored ULL population to produce a pulse, then a short time later open the Q-switch fully to produce a second pulse (which usually depletes what remains of the ULL population).

A single-pulse Q-switched flashlamp-pumped solid-state laser system works by firing the flashlamp, allowing a large inversion to build; then, at a predetermined optimal time where the ULL population reaches maximum (predicted by a model

---

EXAMPLE 6.7   OUTPUT POWER OF A Q-SWITCHED LASER

A ruby laser has the following parameters:

$$R_{HR} = 100\%$$
$$R_{OC} = 85\%$$

Rod Size = 7.5cm long by 9.5mm diameter

Attenuation = 0.02m^{-1}

One could assume the small-signal gain of ruby to be the accepted value of 20m^{-1}, but for better accuracy the gain was determined experimentally using the technique outlined in Section 2.4 (since this laser features an EOM Q-switch that allows adjustment of drive voltage and so, in effect, the EOM Q-switch is used here as a variable intra-cavity loss). Thus, the gain was determined (with the flashlamp operating at maximum input energy) to be 17m^{-1}. The maximum inversion (now the initial inversion for the Q-switched pulse) is computed according to Equation (2.20) to be:

$$\Delta N_i = \frac{g_0}{\sigma_0} = \frac{17m^{-1}}{2.5 \times 10^{-24} m^2} = 6.8 \times 10^{24} m^{-3}$$

Threshold gain is computed in a similar manner as developed in Equation (2.13), but with the inclusion of the insertion loss of the Q-switch in the "open" state as follows:

$$g_{th} = \gamma + \frac{1}{2x} \ln\left(\frac{1}{R_1 R_2 T_{Q-OPEN}^2} \right) = 0.02m^{-1} + \frac{1}{2(0.075m)} \ln\left(\frac{1}{1.0(0.85)(0.944^2)} \right)$$

Thus, threshold gain is found to be 1.87m^{-1}. In a similar manner to that above, the threshold inversion (ΔN_{th}) is found to be 7.49*10^{23}m^{-3}.

The peak number of photons in the cavity may now be computed using Equation (6.20) by substituting $\Delta N = \Delta N_{th}$ (since peak output occurs when inversion reaches the threshold level) and is computed as follows:

$$n_{PEAK} = \frac{1}{2} \Delta N_{th} \ln\left(\frac{\Delta N_{th}}{\Delta N_i} \right) - \frac{1}{2} \Delta N_{th} + \frac{1}{2} \Delta N_i = 2.20 \times 10^{24} m^{-3}$$

presented later in this chapter, or determined experimentally) the Q-switch is pulsed with a voltage equal to the half-wave voltage (or quarter-wave voltage, depending on the design), resulting in the production of a single massive pulse. A practical Q-switched laser, then, requires a time-delay generator and a pulse generator circuit that can apply a voltage pulse to the Pockels cell to fire the laser.

A double-pulse laser system works by applying two pulses to the Q-switch at pre-programmed times (usually after the inversion has built to an optimal level). Unlike

In order to compute peak power, cavity photon lifetime is required. As per Equation (3.50) it is computed as:

$$\tau_{CAVITY} = \frac{1}{\left(\dfrac{c}{n}\right) \times g_{th}} = \frac{1}{\left(\dfrac{c}{1.7}\right) \times 1.87m^{-1}} = 3.03ns$$

Power is then computed according to Equation (6.23):

$$P = (1 - R_{OC})h\upsilon\frac{nV}{\tau_{CAVITY}} = (1 - 0.85)\frac{hc}{694 \times 10^{-9}m}\frac{2.20 \times 10^{24}m^{-3}(0.075m)(\pi 0.00475^2)}{3.03 \times 10^{-9}s}$$

$$= 1.66 \times 10^8 W$$

And energy is calculated according to Equation (6.26):

$$E = \frac{1}{2}(1 - R_{OC})h\upsilon V\Delta N_i$$

$$= \frac{1}{2}(1 - 0.85)\frac{hc}{694 \times 10^{-9}m}(6.8 \times 10^{24})(0.075)(\pi 0.00475^2) = 0.78J$$

It might be noted that one could approximate the duration of this pulse by dividing the energy by the power (both computed above) to yield an approximate pulse length of 5ns. (The actual pulse is 10ns in length but is not a "square" pulse.)

While the energy of the pulse is moderate, the power of this laser is enormous at 166MW and this example nicely illustrates why Q-switching is used. If this same laser is used in "long pulse" mode with a 200μs-wide output pulse, the energy would be the same but the power of the pulse is only 2550W. The physical effect of each type pulse on various materials will be quite different, with the long pulse penetrating deeply into most materials (and hence being more suitable, for example, for welding) while the Q-switched pulse will essentially ablate the surface of most materials it strikes.

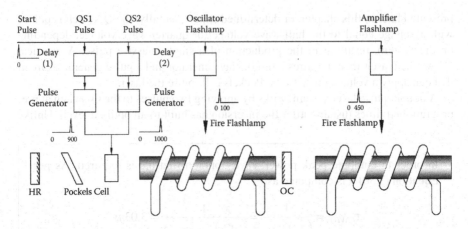

FIGURE 6.13 Configuration and sequencing of a double-pulse laser system (see text).

a single-pulse laser, the first pulse is of a variable voltage—the higher the applied voltage, the higher the transmission of the switch for that pulse. Pulses can vary, logically, from zero volts (in which case the first pulse will be absent) up to the half- or quarter-wave voltage (specific to the implementation of the laser), at which point all of the ULL population will be used. The first pulse will hence use an intermediate voltage such that it does not utilize the entire ULL population available, leaving enough population stored in the amplifier for a second pulse. The second pulse is set for an optimal output (i.e half- or quarter-wave voltage like a single-pulse laser) to utilize any inversion remaining. In this manner two pulses are produced from the same laser amplifier rod, with identical characteristics.

The basic configuration for a typical double-pulse laser is outlined in Figure 6.13. This particular laser is a MOPA (Master Oscillator Power Amplifier) configuration in which a pulse from a smaller laser (the Master Oscillator) is later amplified by a single pass through a power amplifier that is separately pumped. This arrangement is typical of large, high-power lasers. The production of a double pulse begins with the firing signal at $t = 0$. After a short delay the master oscillator lamp is fired, building an inversion in that amplifier; then, at a later time, the amplifier lamp is fired, building an inversion in that amplifier as well. (These times are shown on the figure as "100µs" and "450µs" respectively to correspond with the control settings shown in Figure 6.14, which follows.) Now that an inversion is built, the laser (or

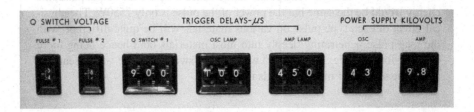

FIGURE 6.14 Controls for a double-pulse laser.

rather the oscillator) Q-switch is opened to produce the first pulse, which is then amplified by the power amplifier and exits the laser. A short time later the Q-switch is opened again, this time at optimal settings to deplete any remaining inversion in the oscillator rod. This second pulse is also amplified and exits the laser.

Figure 6.14 details the controls for a double-pulsed ruby laser used for holography. The basic laser is the same as that outlined in Figure 6.13 (i.e. with a MOPA configuration). The Q-switch parameters (set via the first three controls from the left), which may be varied, include the voltage for each pulse and the timing of the first pulse—it is shown in the figure at 900μs and since the second pulse is fixed at 1ms the pulses are spaced 100μs apart. Other controls (on the right) control the timing (via the trigger delays) and pumping (via the capacitor voltage) of the oscillator and amplifier rods, which affects gain and overall power output.

For the first pulse, the initial inversion starts at the small-signal gain and decreases to a midpoint level. For the second pulse, the initial inversion starts at that midpoint and terminates at the threshold level (where the switch is fully open). In mathematical terms, one equates the energy of each pulse and solves for the midpoint gain. Once computed, the inserted loss of the switch to effect that gain may be computed and finally the midpoint voltage. In reality, solution of the equations is difficult and is easiest to accomplish by using a simple numerical model.

From the model developed in Section 6.6, the power of any given pulse is a function, generically, of the initial (ΔN_i) and final (ΔN_f) inversions (from Equation 6.25) as:

$$P_{PEAK} = (1 - R_{OC})h\upsilon V \frac{1}{\tau_{CAVITY}} \left[\frac{1}{2}\Delta N_f \ln\left(\frac{\Delta N_f}{\Delta N_i}\right) - \frac{1}{2}\Delta N_f + \frac{1}{2}\Delta N_i \right] \quad (6.27)$$

where for single-pulse operation, the initial inversion is usually determined from the small-signal gain of the medium and the final inversion usually determined from the threshold of the laser configuration. For double-pulse operation, the first pulse occurs between the initial inversion (again, computed from the small-signal gain of the amplifier, which for practical purposes most often can be measured experimentally) to a midpoint threshold (effected by a partially open Q-switch) with the power of this first pulse computed as:

$$P_{PEAK} = (1 - R_{OC})h\upsilon V \frac{1}{\tau_{CAVITY}} \left[\frac{1}{2}\Delta N_{MID} \ln\left(\frac{\Delta N_{MID}}{\Delta N_i}\right) - \frac{1}{2}\Delta N_{MID} + \frac{1}{2}\Delta N_i \right] \quad (6.28)$$

Similarly, the second pulse is generated with the initial inversion starting at the midpoint (the inversion left in the amplifier after the first pulse) to the "true" threshold value with the Q-switch set for a "fully open" state. The power of this second pulse is computed as:

$$P_{PEAK} = (1 - R_{OC})h\upsilon V \frac{1}{\tau_{CAVITY}} \left[\frac{1}{2}\Delta N_{TH} \ln\left(\frac{\Delta N_{TH}}{\Delta N_{MID}}\right) - \frac{1}{2}\Delta N_{TH} + \frac{1}{2}\Delta N_{MID} \right] \quad (6.29)$$

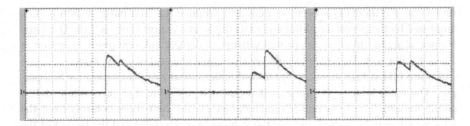

FIGURE 6.15   Double-pulse laser output.

where N_i is the initial inversion, N_{MID} is the midpoint inversion, and N_{TH} is the threshold inversion computed from the threshold gain.

Experimentally, this midpoint could be found by first determining the optimal voltage for the second Q-switch pulse (which should be the half- or quarter-wave voltage for the Q-switch). The midpoint voltage is then set for zero volts, the laser fired, and the output recorded. Based on the results, the midpoint voltage is increased or decreased until equal pulses are produced. For example, if the second pulse was larger than the first, too much ULL population would remain after the first pulse so the voltage on the first pulse would be increased to extract more ULL for that pulse. Eventually, equal pulses are produced.

Actual output from a double-pulse laser with various settings is seen in Figure 6.15. The laser was triggered at 0μs with the first pulse appearing at 900μs after triggering. (Each grid on the oscilloscope screen represents 100μs and the trigger point of 0μs is off to the left of the screen and so is not shown.) The leftmost section shows the results when the first pulse was set too large (meaning the Q-switch voltage for the first pulse was too high). The middle section shows the opposite situation, and the section on the right shows an optimal output where the two pulses are identical.

Using the expressions for pulse power (Equations 6.27 and 6.28), a simple numerical model may be used to determine the optimal setting for the midpoint voltage. In this model, which uses a spreadsheet for simplicity of implementation, three columns are set up to represent the intermediate inversion, the power of the first pulse, and the power of the second pulse. The power of the first pulse is computed as a function of the inversions between the initial value and the midpoint, and the power of the second pulse as a function of the inversions between the midpoint and the threshold value. Midpoint inversions are chosen to range between the initial and threshold values at arbitrarily small increments (with smaller increments increasing the accuracy of the model). Where the pulse energies are seen to be equal, the midpoint inversion is found and can be converted into a value of voltage for the Q-switch (i.e. the midpoint voltage, which becomes the voltage to be applied to the first Q-switch pulse).

Example 6.8 outlines the implementation of the model for an actual laser and simplifications based on simple proportionalities. (For example, since gain is proportional to inversion, one may choose to consider initial, midpoint, and threshold gains instead of inversions—gains that must be computed anyway in the course of analysis of this laser.)

EXAMPLE 6.8   PREDICTING Q-SWITCH SETTINGS FOR A DOUBLE-PULSE LASER

In this example, the settings for a double-pulse ruby laser will be computed to generate pulses of identical energy. The laser is the same one used in Example 6.7 but is used in "double pulse" mode in this case. It has the following parameters:

$$R_{HR} = 100\%$$
$$R_{OC} = 85\%$$
Rod Size = 7.5cm long
Attenuation = 0.02m^{-1}

The same Q-switch as described in Example 6.5 was used (which was imperfectly aligned) and so the effective half-wave voltage (where maximum output was observed) was 5.7kV. This corresponds to an offset of $\phi = 0.3174$ in Equation (6.6).

With a reduced pump power (i.e. with the oscillator flashlamp operated at 70% of maximum energy), the laser was found to oscillate at threshold with an applied Q-switch voltage of 3.4kV. The transmission of the switch at this voltage is hence found by Equation (6.6) to be 77.8%. The small-signal gain of the medium may thus be calculated according to:

$$g_{th} = \gamma + \frac{1}{2x} \ln\left(\frac{1}{R_1 R_2 T_Q^2} \right)$$

With substitution of the parameters (including the Q-switch transmission, denoted T_Q in the formula and squared since it is encountered twice per round trip through the amplifier), the gain is computed to be a modest 4.45m^{-1}.

The threshold gain of the laser is now calculated in a similar manner with the transmission of the Q-switch equal to the maximum transmission (i.e. minus insertion loss) of 0.944 to yield a threshold gain of 1.87m^{-1}.

The inversion densities can now be computed and Equation (6.24) used to compute the power of each pulse, but since we do not require an answer in terms of actual energy per pulse (but rather only an indication of when the pulse energies are equal) a simplification exists by realizing that the pulse of the pulse is proportional as follows:

$$P_{PULSE\_1} \propto g_0 - g_{MID} - g_{MID} \ln\left(\frac{g_0}{g_{MID}} \right)$$

since the inversion for the first pulse begins at ΔN_0 (which is proportional to g_0) and terminates at ΔN_{MID} (which is proportional to g_{MID}). It follows that for the second pulse,

$$P_{PULSE\_2} \propto g_{MID} - g_{TH} - g_{TH} \ln\left(\frac{g_{MID}}{g_{TH}} \right)$$

	A	B	C
1	**Double-Pulse Solution**		
2			
3	g_0	4.45	
4	g_{th}	1.87	
5			
6	g_{mid}	$E_{1stPulse}$	$E_{2ndPulse}$
120	3.00	0.27	0.25
121	3.01	0.26	0.25
122	3.02	0.26	0.25
123	3.03	0.26	0.26
124	3.04	0.25	0.26
125	3.05	0.25	0.27

FIGURE 6.16 Spreadsheet output for equal pulses.

A spreadsheet can then be developed with g_{MID} varying from 4.45m^{-1} to 1.87m^{-1} in increments of 0.01m^{-1}, as outlined in Figure 6.16. As expected, with the midpoint set for threshold gain the first pulse uses all available inversion and the second pulse is nonexistent. Similarly, when the midpoint is set to the initial gain the first pulse is absent and the second pulse uses all available inversion. By inspecting the output of the model, equal pulses are found to exist when $g_{MID} = 3.03$m^{-1} as shown in row 123 of the spreadsheet. Note that the pulse energy shown in columns B and C are not actual energy in Joules since the proportionalities outlined here are used.

This midpoint gain figure can now be used to compute the transmission of the EOM required, as per the method outlined in Example 6.1, as:

$$g_{th} = 3.03m^{-1} = \gamma + \frac{1}{2x} \ln\left(\frac{1}{R_1 R_2 T_Q^2} \right)$$

or, solving for T_Q numerically, 86.5% transmission. Using the relation developed in Examples 6.4 and 6.5, the corresponding Q-switch voltage for this first pulse is then found to be 4.14kV.

Experimentally, the optimal value for double-pulse operation is found to be 4.5kV, in good agreement to this prediction. (The actual methodology used was to set the Q-switch voltages to the values predicted by this model and then refine these by trial-and-error, varying the voltages 100V at a time and firing the laser.) This is actually seen in Figure 6.15, in which the voltages used to produce these three figures were as follows: for the leftmost figure with the first pulse too large 4.7kV and 5.7kV, for the middle figure with the first pulse too small 4.1kV and 5.7kV, and for the final figure with equal pulses 4.5kV and 5.7kV. Although the theoretical half-wave voltage at the lasing wavelength of 694nm is 8.35kV, the actual half-wave voltage is 5.7kV. (Refer to Example 6.5 for details on the alignment of the EOM.)

Depending on the separation of the two pulses and the time at which they are produced, re-pumping of the ULL population in the oscillator is possible between pulses in a double-pulse laser. If, for example, the timing is set such that the pump lamp is still active after production of the first pulse, the ULL will be replenished to some extent (which may be significant if a long separation in time exists between the two pulses). Assuming the solid-state medium chosen has a long ULL lifetime, one could wait until the pump lamp has produced most of its output before producing Q-switched (i.e. fire the pump lamp early and rely on the fact that the amplifier will maintain this population without significant decay until pulses are produced). However, if the pump lamp is still active, inaccuracies will be seen in the model (although, in practice, they are usually overcome with simple manual adjustment of the appropriate Q-switched voltage).

As a final aside, the same laser employed in Example 6.8 was also built in quad-pulse configurations where four individual pulses were possible in a single firing!

6.8 MODELING FLASHLAMP-PUMPED LASERS

As pumping of an amplifier ensues, population builds, and for a Q-switched laser, the switch is opened when maximum inversion exists for optimal pulse power. Determining the timing of this pulse is critical to efficient operation and so a model is presented here to determine that optimal time.

Consider first the intensity of a flashlamp pulse used to pump a solid-state laser as depicted in Figure 6.17. An incorrect assumption would be that maximum inversion is achieved when the flashlamp reaches maximum intensity since the ULL

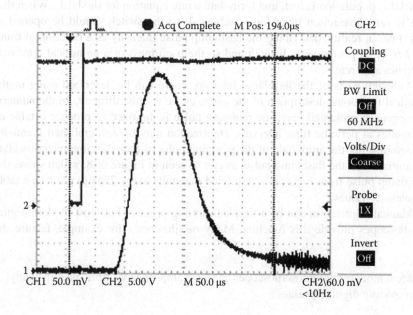

FIGURE 6.17 Time dependency of the intensity of a flashlamp-pumping pulse.

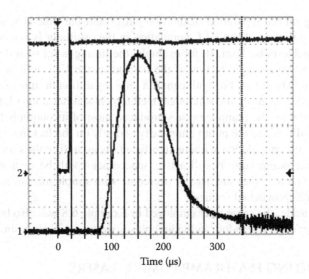

FIGURE 6.18 The flashlamp-pumping pulse digitized every 25µs.

acts as an integrator storing atomic population, at least until decay (characterized by the lifetime of the level) becomes significant enough to depopulate the level. In theory, a calculus-based solution is both simple and elegant: describe the intensity of the flashlamp pulse with a mathematical function (which describes the rate at which the ULL population grows), use the exponential decay function to describe how ULL population is lost, and formulate a rate equation for the ULL. When this rate is zero, maximum inversion is achieved, so the switch should be opened at this time. In reality, describing the flashlamp pulse intensity as a function of time, i.e. $I(t)$, often proves very difficult and so the usefulness of a numerical approach becomes apparent.

Although parts of the flashlamp intensity curve can be described using mathematical functions, description of the entire curve is quite difficult. In the numerical approach presented here, the pumping pulse is digitized to produce a table of intensities at periodic time intervals. Digitization may be accomplished manually by determining the amplitude of the pumping pulse at periodic time intervals after the initiation of the flashlamp pulse, as can be seen in Figure 6.18, which shows the flashlamp pulse from Figure 6.17 digitized manually every 25µs to produce a table of intensity values.

Manual digitization can be a time-consuming process, but most modern digital oscilloscopes provide this function. Many oscilloscopes, for example, feature the

Most modern digital oscilloscopes can save amplitude vs. time data, instantly providing digitized values.

ability to print both a screen shot (as an image) and a table of values to a memory key. The table of values can conveniently be saved in the CSV (Comma Separated Value) file format, which can be opened by most spreadsheet programs. Often, two columns in the file represent time in seconds and amplitude in volts. Very high resolution is also possible with many scopes. (With a timebase of 50μs/division selected, for example, the oscilloscope used to capture the intensity in the previous two figures recorded intensity values at 0.2 μs intervals—the result being that ten divisions of the screen results in 2500 data points.)

Regardless of how accomplished, the result of the digitization is a series of intensity values. While these could theoretically be calibrated in real units (for example, W/m^2), arbitrary units are often used since the goal of this model is simply to predict when to fire the Q-switch, not the actual inversion at that point (although that, too, will be explored later).

The model then proceeds by computing the ULL population at any time t that consists of the number of ions pumped in during this time interval as well as the amount remaining from the previous time interval, now decayed exponentially according to:

$$N(t) = N(t-1)e^{\frac{-\Delta t}{\tau}} \qquad (6.30)$$

where $N(t)$ is the current atomic population and $N(t-1)$ the population at the previous time interval (Δt).

Since this is an iterative process, a spreadsheet greatly simplifies the model. One could represent the contribution of each time interval (i.e. the "leftover" population from previous quantities of atoms) to the total population at any time t by computing the fraction of atoms remaining in the ULL from each previous time interval. These populations are then summed to yield the total population remaining:

$$N(t) = N_{PUMPED} + N_{PUMPED}(t-1)e^{\frac{-\Delta t}{\tau}} + N_{PUMPED}(t-2)e^{\frac{-2\Delta t}{\tau}} \qquad (6.31)$$

As an example, consider from Figure 6.18 the population at 100μs to be:

$$N(100\mu s) = N_{PUMPED}(100\mu s) + N_{PUMPED}(75\mu s)e^{\frac{-25\mu s}{230\mu s}} + N_{PUMPED}(50\mu s)e^{\frac{-50\mu s}{230\mu s}} \qquad (6.32)$$

τ was chosen in this example for the Nd:YAG laser (230μs). Implementing this as a spreadsheet would require multiple columns, each representing a population of atoms pumped into the system at each discrete time interval (every 25μs in this example) and each decaying in time. At any time t, then, the population at the ULL is simply the sum of all of these terms, which in a spreadsheet would be the sum of each row. With a small time interval, the spreadsheet will be quite large!

An easy simplification to the model would be a recursive approach where the population at any time could be represented based on the population at the last time interval. For example, Equation (6.32) could be simplified to:

$$N(100\mu s) = N_{PUMPED}(100\mu s) + N(75\mu s)e^{\frac{-25\mu s}{230\mu s}} \qquad (6.33)$$

This makes the formulation of a spreadsheet much simpler, especially when many data points are involved (i.e. where Δt is small, as required for accuracy with a numerical model such as this). To apply this to a spreadsheet, one need only develop the formula in one cell, so that when copied to the next row, the cell references will increment accordingly.

So, the actual population at any time interval t is expressed generically as:

$$N(t) = N_{PUMPED}(t) + N(t - \Delta t)e^{\frac{-\Delta t}{\tau}} \qquad (6.34)$$

where N_{PUMPED} represents the number of atoms pumped to the ULL in this time interval (which is proportional to the pump intensity at this time interval) and $N(t - \Delta t)$ is the population of the ULL during the last time interval (Δt).

EXAMPLE 6.9   CALCULATING THE TIME FOR PEAK INVERSION

Consider a YAG laser, for which Figure 6.17 provides the flashlamp intensity. The intensity of the flashlamp pulse was digitized at intervals of 25µs in arbitrary units of mV since the photosensor used to monitor the intensity of the flashlamp produces an output ranging from 0 to 500mV where this voltage is proportional to intensity.

Although triggered at $t = 0$, from Figure 6.17 it is evident that light output from the flashlamp does not begin until approximately 75µs, peaking at about 150µs. At 100µs, the flashlamp intensity was found to be 115mV and so $N(100\mu s) = 115$ (where 115 is in arbitrary units). At 125µs, the flashlamp intensity is found to be 282, so the total population is calculated according to Equation (6.34) to be:

$$N(125\mu s) = N_{PUMPED} + N(100\mu s)e^{\frac{-t}{\tau}} = 282 + 115e^{\frac{-25\mu s}{230\mu s}} = 282 + 103.2 = 385.2$$

At 150µs, the flashlamp intensity is found to be 330, so the total population is now 330 plus the previous population of 385.2 decayed, again, by 25µs.

A spreadsheet may be employed to compute the population at any time interval. By copying the basic formula into successive rows cell references are incremented for each new row.

	A	B	C	D
1	Inversion in a Flashlamp-pumped Laser			
2				
3	Tau(μs)	230		
4	Δt (μs)	25		
5				
6	t	FLP Power	Null	ΔN/Gain
7	(us)	(mV)	Decayed	(theoretical)
8	0	0	0	0.0
9	25	0	0	0.0
10	50	0	0	0.0
11	75	0	0	0.0
12	100	115	0	115.0
13	125	282	103.1553884	385.2
14	150	330	345.485684	675.5
15	175	305	605.9129397	910.9
16	200	217	817.091983	1034.1
17	225	120	927.5840009	1047.6
18	250	70	939.6863865	1009.7

FIGURE 6.19 Spreadsheet.

The spreadsheet is outlined in Figure 6.19, in which digitized flashlamp intensity data is entered into columns A and B. Column C represents the atomic population from the last time interval and column D the total population at this time. The formula for cell C14, for example, is "=D13*EXP(-$B$4/$B$3),", which will compute the total population of the ULL, decayed by 25μs. Note that the time interval and ULL lifetime are stored at the top of the spreadsheet and referred to by absolute reference allowing rapid "What if?" scenarios to be investigated. (For example, "What if vanadate, which has a different ULL lifetime, was used instead of YAG?")

The formula for cell D14 is "=B14+C14," which is the sum of the current pump intensity plus the amount left from the last time interval, decayed by 25μs.

So, cells in columns C and D for each successive row compute the current population of the ULL from the last time interval decayed by Δt, to yield the "leftover" population in the amplifier, summed with the population currently pumped into the system (proportional to intensity of the flashlamp). In the example provided here, a relatively crude time interval of 25μs was employed, but where the data was taken directly from the oscilloscope for this same example, intensity values were digitized every 0.2μs (meaning 2500 data points for a 500μs capture, as shown in Figure 6.17). Such a spreadsheet-based model will hence be 2500 rows in length (illustrating the value of using a spreadsheet).

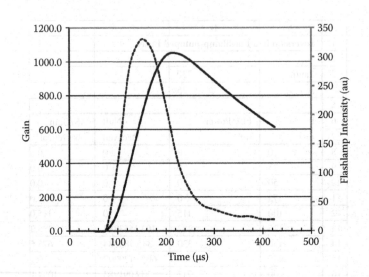

FIGURE 6.20 Inversion as a function of time.

The output from the model, using a small time interval of 0.2µs, is seen in Figure 6.20, plotted along with the flashlamp intensity (as a dotted line). The model reveals that the peak inversion occurs at a significantly delayed time from the peak flashlamp intensity and so setting the Q-switch opening to coincide with the peak flashlamp output will yield a very low output power. (At low pump intensities, output at that point will be completely absent.)

Since the model produces an output that indicates, at least proportionally, how large the inversion is at any time t, knowing where threshold occurs allows the model to be calibrated in real units of inversion or gain (which is proportional to inversion), so parameters such as actual maximum gain may be computed by combining the output from the model with a few experimental measurements. Experimentally, one only needs to determine the Q-switch delay that allows oscillation to begin: the delay is set to zero, the laser fired repeatedly, and the delay increased until an output pulse appears. At that point, threshold inversion is reached and the threshold gain may be computed in the normal manner (as outlined in Chapter 2) knowing the optical parameters of the laser.

To determine the maximum inversion (or maximum gain), one may simply scale the peak gain from the model with the threshold gain already determined.

EXAMPLE 6.10   CALIBRATING THE MODEL

For the laser in Example 6.9, the Q-switch delay was varied until output was seen, which occurred at 140µs. This means that the inversion has built, at this time, to the threshold level, which is computed by knowing the optical parameters of this laser:

$R_{HR} = 100\%$
$R_{OC} = 90\%$
Rod Size = 115mm long
Attenuation = $0.1m^{-1}$
Q-Switch transmission (open) = 96%

The threshold gain is hence computed to be $0.913m^{-1}$.

In the output from the model, the inversion at 140μs was found numerically to be 618 arbitrary units. The actual peak inversion is seen to occur at 211.4μs and numerically is 1051 arbitrary units. The peak gain of the laser can then be found with a simple ratio as:

$$g_0 = g_{th} \frac{1051}{618} = 1.55m^{-1}$$

Of course, by knowing the initial and threshold inversions for a pulse one may use the previous model to compute the power and energy of the resulting pulse. In the case of an arbitrary firing time, the initial inversion would be the inversion at that point in time (usually the peak inversion would be used for optimal pulse power).

6.9   REPETITIVELY PULSED Q-SWITCHED LASERS

In a laser pumped by a CW source of constant pump intensity (e.g. a diode-pumped solid-state laser or DPSS), and with the Q-switch closed, inversion builds until it reaches an equilibrium value. This value can be quite large (certainly much larger than the threshold value) and depends on both the rate of pumping and the ULL lifetime. A large pumping rate will increase the ULL population, but this population is lost through natural exponential decay from the ULL (characterized by the lifetime of the ULL state, τ_{ULL}). When the rate of pumping equals the rate of decay, the ULL population will have reached this equilibrium value. This can be demonstrated visually by modifying the model from the previous section to use a constant pump intensity instead of one that varies in time as the output from a flashlamp does.

Many modern DPSS lasers are operated in a mode in which the pump source (a diode laser) remains on with constant intensity and the Q-switch is opened repeatedly to produce a stream of pulses. For ultimate pulse energy, one could wait for the inversion to build to the ultimate equilibrium level, many times longer than the ULL lifetime, before opening the switch; however, this will result in a relatively slow pulse rate (< 1KHz), which may be unsuitable for many applications where pulse rates of well over 10KHz are desired. Operation at such fast rates will result in a pulse energy lower than the maximum possible, but it can be easily predicted.

For any given constant pump intensity, the fraction of maximum inversion reached is a function solely of the ULL lifetime and the pump intensity. After a time period of

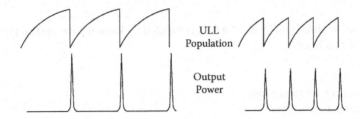

FIGURE 6.21 N_{ULL} buildup at low and high repetition rates.

twice the lifetime of the ULL (τ_{ULL}), the inversion reaches 85% of the final value, and within four time periods 98% of the final value is reached. Assuming a long enough time period (for example, four time periods) elapses between pulses, opening the Q-switch at this point results in essentially the largest output pulse possible; however, the purpose of this model is to determine when to open the Q-switch to produce a known energy (for example, 90% of this maximum), as this will set the maximum practical pulse frequency for a given application. In a previous section a model was presented allowing prediction of pulse energy as a function of inversion so that if we know the inversion (as well as a few key parameters of the laser), we can predict actual output from the laser.

Figure 6.21 illustrates the effect of Q-switch rate on pulse power for a continuously pumped laser where this rate is fast enough to prohibit the ULL achieving a population approaching the maximum value.

In a practical, continuously pumped, repetitively-pulsed laser, then, pulse energy is inversely proportional to pulse repetition rate: As this rate increases, the ULL is given less time to build a large population and so pulse energy decreases.

Where a continuous pump source is employed, predicting inversion as a function of time (and hence pulse energy for a given repetition rate) is somewhat intuitive, from a mathematical viewpoint. The rate equation is simple: population of the ULL grows by a constant term and decays exponentially as characterized by the ULL lifetime. The maximum value achieved obviously occurs at an infinitely long time period—practically, though, we reach a level of over 99% of the maximum after a period of five times the ULL lifetime—but the inversion (and then the available pulse energy) at any arbitrary time t may be found according to:

$$N(t) = N_{MAX}\left(1 - e^{\frac{-t}{\tau}}\right)\tag{6.35}$$

where N_{MAX} is the maximum inversion that will be achieved (after a very long time) and is mathematically proportional to the total amount of energy pumped into the system in a time of τ seconds—for a constant pump intensity this value would be that intensity multiplied by τ.

Equation (6.35) can be solved directly and plotted to reveal the inversion as a function of time, as shown in Figure 6.22 (where τ is the time constant of the amplifier medium, 230µs for YAG in the example illustrated here). As time

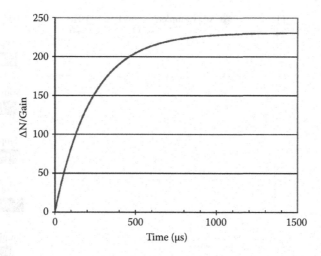

FIGURE 6.22   Inversion/gain buildup in a continuously pumped amplifier.

increases, the inversion asymptotically approaches the maximum level but practically after five time constants (i.e. 5τ) the inversion builds to over 98% of the final value.

6.10   GIANT FIRST PULSE

Many processes, such as drilling and engraving, employ a continuously pumped Q-switch laser as described previously—the short Q-switched laser pulse producing the desirable effect of ablating material without significantly heating the substrate below. In these cases the laser is turned off by inhibiting the Q-switch, the beam is positioned over a target area, and the laser is fired repeatedly. When a hole or engraved line is complete, the laser is again inhibited and the beam is repositioned. In these applications an effect occurs by which the first of the series of pulses produced is much larger than the rest, known as the *giant first pulse*. As illustrated in Figure 6.23, in which the top trace represents opening of the switch and the bottom trace the resulting output pulse, when the Q-switch is inhibited for a length of time (for example, during beam positioning), a much larger than normal inversion builds. The first pulse emitted is thus considerably more powerful than the rest that follow. This can result in an undesirable "crater" forming at the beginning of a cut line.

The effect is most pronounced when repetition rates are large. Where a slow rate is used (and hence the time between pulses is greater by many times than the ULL lifetime), the inversion between each pulse builds to a value close to the maximum regardless and so in effect each pulse is a "giant" pulse; however, with a fast repetition rate, the inversion between each pulse is not afforded the opportunity to build to even close to the maximum level, resulting in much smaller pulse powers. This is illustrated in Figure 6.23 as well, where the pulse rate is 5KHz and so between each pulse inversion builds only for 200µs. With a YAG laser (with an ULL lifetime of 230µs) this means inversion builds to only 58% of the maximum value according to

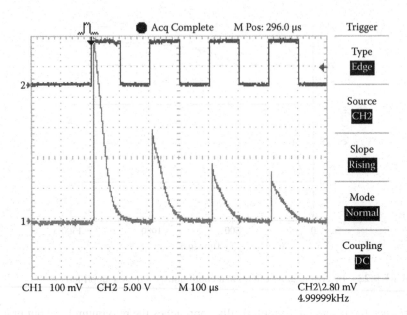

FIGURE 6.23 A giant first pulse.

Equation (6.35). Before the first pulse (lower trace) appears, the laser is idle for over 2ms and so the inversion for that pulse is essentially 100% of the maximum at that pump rate.

The size of the effect is a function of ULL lifetime. The longer the ULL lifetime, the larger this first pulse becomes. Vanadate (Nd:YVO$_4$), for example, has a shorter ULL lifetime than Nd:YAG, so the effect is less pronounced. By using the model already developed, one can predict the magnitude of the first pulse as opposed to other pulses in the series by knowing the time for which the ULL is allowed to build population: for the first pulse the laser is off long enough that the population builds essentially to the terminal equilibrium level, and for pulses that follow in the series the time to build population is the inverse of the pulse rate. The model does not take into account residual populations remaining after production of a pulse (since it is possible that the Q-switch closes before allowing the population of the rod to be reduced to the lowest value possible). This is evident in Figure 6.23, in which the giant pulse phenomenon affects not only the first pulse but the one that follows it.

There are numerous methods to overcome the "giant first pulse" problem. One such method called *First Pulse Suppression* (FPS) involves elimination or reduction of the first pulse by causing the inversion to "bleed down" slowly instead of being used for the production of a pulse. With an EOM switch, a slow ramp of the Q-switch voltage (as opposed to the fast pulse normally used to generate a Q-switched pulse) will cause the laser to oscillate at low power levels, which in turn depletes the stored inversion. With an AOM switch, a similar slowly decreasing ramp of the RF drive power will gradually increase the Q of the cavity and deplete the inversion. These schemes usually result in a missing first pulse. While the first pulse can easily be eliminated altogether with this scheme, it can also simply be reduced by opening the Q-switch more and more for each time the switch is opened during production of a series of pulses. (After a while, the switch will fully open for each pulse as expected.) With careful adjustment, this scheme can be used to ensure all pulses are of equivalent power. Other schemes involve modulation of the pump diode current rather than the Q-switch drive to prevent the massive inversion from building in the first place (in other words, turn the pump diodes off or to a low level between pulses).

6.11 ULTRA-FAST PULSES: MODELOCKING

Modelocking is a technique for producing ultra-fast pulses far shorter than a Q-switched laser can produce. While the basic, original modelocking technique can be used to produce a series of repetitive pulses, the technique can be used in a variety of ways to produce picoseconds or even femtosecond pulses.

The basic idea of modelocking is to employ an amplifier with a large gain band-width, which under normal laser conditions (i.e. in a standing-wave laser) would sup-port a large number of longitudinal modes each of a slightly different frequency. In a modelocked laser, however, a phase relationship is established for a moment in time between these modes such that they are all in-phase and as such an enormous ampli-tude results. This is illustrated in Figure 6.24, in which many modes are summed together to yield the pulse shown (with the envelope also shown on the figure as a dashed line). Quickly, though, these modes interfere with each other (since phase cannot be kept constant between waves of different frequencies) and the pulse ampli-tude falls rapidly. While in many lasers an amplifier with a wide gain bandwidth is a detracting feature (since it makes production of highly coherent radiation difficult, often requiring the reduction of oscillating modes), in a modelocked laser it is not only desirable but required, and the wider the bandwidth, the shorter the pulse may potentially be.

Modelocking a laser requires an optical switch, generally a Q-switch, which opens only once during a round trip of photons in the laser cavity. In this manner, opening of the switch causes the phase of all modes to lock together in time. By allowing

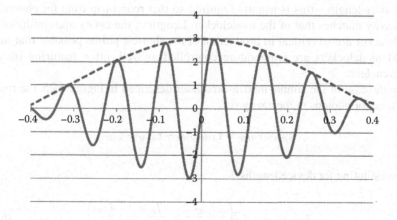

FIGURE 6.24 Phase in a modelocked pulse.

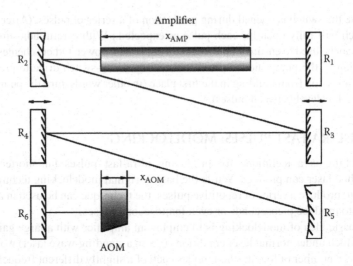

FIGURE 6.25 A modelocked laser.

the switch to open only once per round trip, longitudinal modes are prevented from building in the cavity. Physically, only a small packet of radiation exists in the cavity at any time, continually traversing through the amplifier.

A simple modelocked laser is shown schematically in Figure 6.25 (in which mirrors R_1 and R_6 are the primary cavity mirrors, one is the HR and one the OC, and the other mirrors are all high reflectors in this folded-cavity design). A standing-wave AOM, used as the modelocker, is placed immediately adjacent to cavity mirror R_6, which necessitates that the AOM open only once per round trip, simplifying the timing. Since AOMs designed for modelocking are often fixed-frequency devices with a small bandwidth (and so the rate at which it opens cannot be varied), mirrors R_3 and R_4 are movable, allowing the dimensions of the cavity to be varied to match the AOM rate. The folding mirrors serve no purpose other than to extend the cavity length—this is usually required so that round-trip time for photons in the cavity matches that of the modelocker. Length of the cavity and timing of the modelocker are so critical to production of the shortest pulses possible that many AOM modelockers are temperature controlled as well, often featuring integral TEC coolers.

In the case of the simple modelocked laser described in Figure 6.25, the round-trip time for photons in the cavity is:

$$\tau_{ROUND\_TRIP} = \tau_{AMPLIFIER} + \tau_{AOM} + \tau_{AIR} \qquad (6.36)$$

Or, substituting for device lengths

$$\tau_{ROUND\_TRIP} = 2\left(\frac{l_{AMP}}{c/n_{AMP}} + \frac{l_{AOM}}{c/n_{AOM}} + \frac{l_{AIR}}{c} \right) \qquad (6.37)$$

with the lengths multiplied by two, since we are examining round-trip time. The resulting pulse rate (in Hz) is simply the inverse of the round trip. This laser shown is quite simple and many modelocked lasers feature much more complex designs with numerous optical elements.

EXAMPLE 6.11   MODELOCKING RATE AND LASER SIZE

An AOM modelocker is to be used in a laser configured as per Figure 6.25. The AOM operates at a fixed frequency of 50MHz. Assuming the amplifier is a YAG with a rod length of 50mm length (and an index of refraction of 1.82) and the AOM has a path length of 25mm (and uses SiO_2 with a refractive index of 1.53), calculate the length of the path in "air" to properly modelock this laser.

At a switching rate of 50MHz, the round-trip time of photons in the cavity is 20ns.

Rearranging Equation (6.37) to solve for the path of air within the laser cavity,

$$\frac{l_{AIR}}{c} = \frac{1}{2}\tau_{ROUND\_TRIP} - \frac{l_{AMP}}{c/n_{AMP}} - \frac{l_{AOM}}{c/n_{AOM}} = \frac{1}{2}(20ns) - \frac{0.05m}{c/1.82} - \frac{0.025m}{c/1.53}$$

renders an answer of 2.87m. This is a long cavity path, which explains the need, practically, for a folded cavity arrangement. Note that the chosen AOM specifies the frequency as 50MHz +/- 150KHz, which is a very tight tolerance. The resulting "air path" length will have a tolerance of only three or four mm (hence the necessity for moveable mirrors).

As with Q-switching, the optical switch can be active or passive. The most common active switch is an AOM switch designed specifically for modelocking in which a standing wave is set up within the AOM. The use of passive switches (saturable absorbers) is also possible, with one approach being a combination mirror/saturable absorber known as a SEmiconductor Saturable Absorber Mirror (SESAM).

As described, a modelocked laser will produce a series of repetitive pulses with a short pulse emitted from the OC (either R_1 or R_6) at every round trip. The losses incurred by the OC limit the power of these pulses, so a cavity dumping technique is often used to produce a pulse on demand. In such an arrangement, a modelocked pulse is developed in a laser with low cavity loss (i.e. without an OC) and allowed to circulate continually. The pulse builds in intensity as it circulates until reaching a very large power. When a pulse is desired, intra-cavity power built up inside the optical cavity is "dumped," or deflected, to form the output beam by a "pulse picker," often another AOM or EOM device.

Even with the generation of larger pulses by using a "pulse picker," the energy of ultrashort pulses is usually quite low (although peak power is extremely high). To obtain significant output energy requires a special amplification technique since the

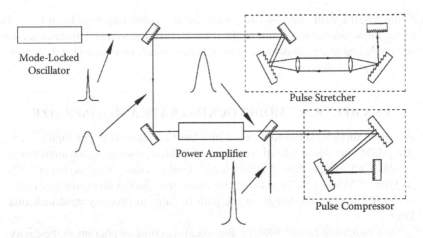

FIGURE 6.26 Chirped-pulse amplification.

power levels of such pulses (which even for a small oscillator can reach beyond the gigawatt level) can easily exceed the damage threshold of almost all known optical materials and coatings. The technique of *chirped pulse amplification* is often used with pulses of enormous power. The process, outlined in Figure 6.26, resembles the MOPA approach outlined in Section 6.7, but the ultrashort pulse is stretched (which, while keeping the energy the same, lowers the power to a manageable level), amplified by a single-pass through one or more amplifiers, and compressed again to the original width.

The key, then, is stretching and compressing the ultrashort pulse. In an example laser of this type, a 100fs pulse from an ultrafast modelocked laser is stretched by many orders of magnitude so that the resulting pulse is 1ns in length. This longer pulse is now amplified and compressed again to the original length of 100fs. In stretching the pulse, various wavelength components of the ultrafast pulse are reflected from several gratings (although occasionally prisms are used) where shorter wavelength components travel further than longer wavelength components so that after passing through this pulse stretcher the pulse is effectively lengthened in time. A typical pulse stretcher is shown in Figure 6.27. In the figure, the incoming

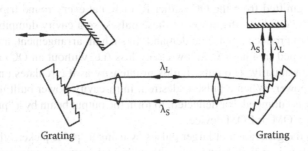

FIGURE 6.27 A pulse stretcher.

ultrafast laser beam (coming from the left) is split by the first grating into long (λ_L) and short (λ_s) wavelengths. Tracing the total path of each wavelength as it strikes the second grating and the mirror, it can be seen that the longer wavelengths incur a shorter path than the shorter wavelengths. The reflected beam now returns through the same path, exiting to the left again with a beamsplitter (not shown) separating the incoming and outgoing beams from the pulse stretcher. A variety of optical configurations are possible for a pulse stretcher, some employing gratings and others utilizing prisms or fiber.

After stretching, the pulse is now "chirped" with long wavelengths appearing first and shorter wavelengths following—like the chirp of a bird that starts with a low frequency and increases to a higher one as the chirp continues. The lengthened pulse has a much lower power which can be safely amplified. Pulse compression is then performed in a similar manner to pulse stretching. Two gratings can be used, again, but this time configured so that shorter wavelengths incur a longer path so that long and short wavelength components exit the compressor at the same time.

Measurement of such a short pulse duration presents a unique problem, since optical detectors do not exist that operate at these speeds, so the *autocorrelation* technique is often used, in which an incoming pulse is split into two identical components, a variable time delay imparted in one beam by means of a moveable mirror and the two beams recombined in a non-linear crystal, where an interaction between the two will occur (non-linear processes are covered in the next chapter). The width of the pulse is then determined as a function of difference in beam length between the two.

7 Non-Linear Optics

No discussion of modern laser technology would be complete without at least a cursory glance at non-linear optics and the production of harmonic radiation. The tendency of some optical materials to behave in a non-linear manner (and in doing so, give rise to the production of high-order harmonics of incident radiation at double, triple, or even higher multiples of the original frequency) is responsible, for example, for the production of green 532nm radiation from the 1064nm YAG laser (the most common example). The technique is somewhat generic, allowing, for example, mixing of various laser wavelengths to produce new wavelengths, and is one reason why solid-state lasers are so popular today. While almost all solid-state lasers oscillate in the IR or near-IR regions of the spectrum, harmonic production opens the door to production of quality laser radiation in the visible and UV regions. Owing to the relatively high efficiencies of conversion now available and the relative simplicity and robustness of solid-state laser systems, frequency-doubled and -tripled lasers now routinely replace gas lasers in a variety of applications.

7.1 ORIGINS OF NON-LINEAR EFFECTS

The non-linear effect is easily understood when the nature of light in a medium is considered. Normally, atoms possess space-charge neutrality—with a strong positive charge in the center and the negative charge distributed evenly in the electron cloud surrounding it; there is no "positive" or "negative" end. As shown in Figure 7.1, as radiation passes through such a medium it is manifested as a wave with an electric field traveling through the medium at a speed of c/n, where c is the speed of light in a vacuum and n the refractive index of the medium. As the wave travels through the medium it induces a macroscopic charge polarization in which the negative charge of the atom (i.e. the electron cloud) is skewed to one side. The resulting atom indeed has a "positive" and "negative" end (best pictured, perhaps, as an egg shape where the negative charge moves toward the small end).

Normally, the amount of polarization is linearly proportional to the magnitude of the electric field applied according to the relationship:

$$P = aE \qquad (7.1)$$

where a is the constant of polarizability and E the electric field vector of the wave. This linearity exists, though, only at low intensities, much like the familiar Hooke's law, which describes the extension of a spring with applied force; when large forces are applied (or, in the case of polarization, a large electric field is applied), non-linearity is observed.

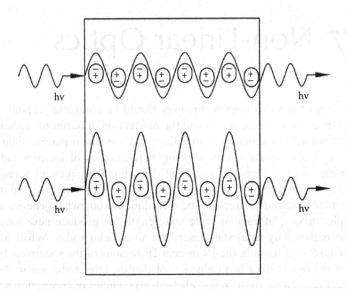

FIGURE 7.1 Charge polarization at low and high intensities.

When a large enough electric field is applied, the polarization is seen to behave according to:

$$P = a_1E + a_2E^2 + \ldots \qquad (7.2)$$

where a_1 is the linear term and a_2 the second-harmonic term. The mathematical basis for production of second-harmonic radiation is easy to understand if one expands the basic polarization equation (Equation 7.2) to incorporate the equation for the oscillating electric field of a wave as:

$$P = a_1E_0 \cos(\omega t) + a_2E_0^2\cos^2(\omega t) + \ldots \qquad (7.3)$$

where ω is the angular frequency (equal to $2\pi f$). Now we may substitute for the second harmonic term using a common identity:

$$\cos^2(\omega t) = \frac{1}{2} + \frac{1}{2}\cos(2\omega t) \qquad (7.4)$$

to yield an equation for polarization as:

$$P = a_1E_0 \cos(\omega t) + \frac{1}{2}a_2E_0^2\cos(2\omega t) + \frac{1}{2}a_2E_0^2 + \ldots \qquad (7.5)$$

Non-linear behavior is observed in springs under large forces that do not follow Hooke's law (which is a first-order approximation regardless).

The first term represents linear polarization (at the fundamental frequency ω), while the second term represents the production of radiation at the second harmonic frequency of 2ω. The last term is a constant DC voltage produced across a crystal that is undergoing non-linear processes—it is a rectification similar to that of an AC signal rectified to DC; it appears as a small voltage across the crystal.

The linear coefficient a_1 is a function solely of the index of refraction of the material according to:

$$a_1 = \varepsilon_0 (n^2 - 1) \qquad (7.6)$$

where ε_0 is the permittivity of free space. The second harmonic coefficient, a_2, is a property of the material itself and varies.

It may be asked, "Why can't we observe non-linear effects in everyday life?" One would expect, for example, that while observing a red source through a quartz glass plate, violet would be seen as well (which it is not). The answer lies in both the magnitude of the non-linear coefficients of the materials used (covered in this section) and in the need to match the phase of the harmonic radiation produced to prevent destructive interference (discussed in the next section).

While a_1 is a large term numerically, a_2 is quite small for all practical materials. Still, given that the second harmonic term is multiplied by the intensity squared (i.e. E_0^2), at large enough intensities efficiency improves, making this scheme practical.

First, we consider available materials. Perhaps the most common non-linear material is potassium titanyl phosphate (KTP). Let's examine the use of KTP as a second harmonic generator (SHG) in doubling the output of a 1064nm YAG laser to 532nm. KTP has an index of refraction of 1.7, and so, according to Equation (7.6), the a_1 term is, numerically, $1.67*10^{-11}$. The second harmonic term, a_2, has an approximate value of $1*10^{-22}$. Now let's calculate the value of the linear and second harmonic terms using Equation (7.5).

The linear term has an amplitude of:

$$P_{LINEAR} = a_1 E \qquad (7.7)$$

and the second harmonic term,

$$P_{HARMONIC} = a_2 E^2 \qquad (7.8)$$

So, by substituting various values for E_0 we can compute the expected conversion efficiency as a ratio of these two amplitudes. This will at least give us a ballpark figure of the intensities required for efficient harmonic generation. We begin with a value of 100V/cm, comparable to the intensity of very bright light (sunlight has an electric field of about 10V/cm), and increase the electric field by several orders

Potassium titanyl phosphate (KTP) is one of the most common non-linear optical materials, but susceptibility to damage limits usage to low- and mid-powered lasers.

TABLE 7.1

An Illustration of SHG Efficiency

E_0 (V/cm)	P_{LINEAR}	$P_{HARMONIC}$	Ratio
100	$1.7*10^{-9}$	$1*10^{-18}$	$6*10^{-10}$
10,000	$1.7*10^{-7}$	$1*10^{-14}$	$6*10^{-8}$
$1*10^6$	$1.7*10^{-5}$	$1*10^{-10}$	$6*10^{-6}$
$1*10^8$	$1.7*10^{-3}$	$1*10^{-6}$	$6*10^{-4}$
$1*10^{10}$	$1.7*10^{-1}$	$1*10^{-2}$	0.06

of magnitude at a time until reaching the value comparable to that of a very tightly focused powerful laser beam (Table 7.1).

SHG conversion efficiency, then, varies with intensity, becoming more efficient as intensity rises. It is therefore advantageous to operate an SHG at the maximum possible intensity (but below the damage threshold of the material). This also explains why non-linear effects are not seen in "everyday" situations where the intensities are low but require the enormous intensities of a focused laser beam: at normal, everyday intensities, the amount of harmonic radiation produced is simply not detectable.

Aside from the enormous intensities required, a non-linear material must be capable of phase matching both the fundamental (original) and second-harmonic component.

7.2 PHASE MATCHING

Once a harmonic is produced in a crystal, like all waves, it is subject to interference (both constructive and destructive) as it travels. Assuming the harmonic is not in phase with the fundamental, destructive interference results in annihilation of the weaker harmonic.

As seen in Figure 7.2, where two generic waves of different wavelength are traveling through a medium, the electric fields of these waves will at some points coincide and hence be in phase, while at other points be out of phase and cancellation will occur. In the case of the non-linear process of second-harmonic generation, one wave will be exactly twice the frequency of the other and we would expect that they would stay in phase forever, as illustrated in Figure 7.3

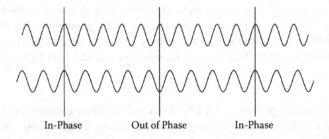

In-Phase   Out of Phase   In-Phase

FIGURE 7.2 Phase matching.

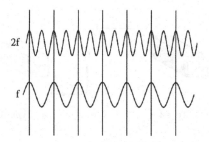

FIGURE 7.3   Phase matching of a second harmonic.

In order for the fundamental and second-harmonic waves to stay in phase with each other, these two components (at different wavelengths) must encounter the exact same index of refraction. The issue here is the behavior of all real optics materials: the index of refraction is not constant and varies with wavelength. The second harmonic, then, is at a shorter wavelength than the fundamental and so will experience a higher index of refraction. This will rapidly cause the harmonic to become out of phase with the fundamental. What is required is a material with the same index of refraction at two wavelengths. This is possible in real materials only if the two wavelengths are at different polarizations, in which case a birefringent material does exactly this. A birefringent material offers two different indices of refraction to two polarizations of light, which at the outset does not help our dilemma—it would still not provide the same index of refraction for both components. However, as the crystal is rotated the index of refraction presented to each polarized component changes. The second harmonic, for example, is initially presented with a higher index of refraction but as the crystal rotates that value decreases. Similarly, the index of refraction for the fundamental wave increases as the crystal rotates and so at some point the indices of refraction will be exactly the same for each component (the fundamental and the harmonic) and we have achieved phase matching.

Phase matching requires precise alignment of the angle of the crystal, which requires great mechanical stability. The actual angle is determined by the cut of the crystal and is often provided by the manufacturer (although it may be determined experimentally by rotating the crystal and monitoring the intensity of the harmonic component produced). Matching in this manner is called *critical phase matching*, reflecting the fact that the angle is a precise parameter and the technique is sensitive to misalignment. Correct choice of angle may allow operation at room or any other temperature. (Refer to Chapter 5 for a description of practical DPSS systems, some of which feature the pump diode and the harmonic crystal operating at the same temperature. It would be advantageous in that case to choose a crystal that phase matches at the same temperature as that required for the pump diode wavelength adjustment.)

Non-critical phase matching may also be achieved by "tuning" the temperature of the non-linear material until phase matching occurs. (Both indices of refraction, one for each polarization, are sensitive to temperature but change at different rates.) In such a scheme the non-linear crystal is often operated at a temperature above room temperature and is controlled by an oven. Careful adjustment of temperature allows phase matching to be achieved, but temperature must be controlled tightly for efficiency. This is the preferred method for most lasers, since it offers very fine control.

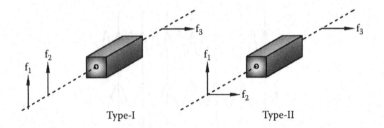

Type-I   Type-II

FIGURE 7.4   Type I and II phase matching.

Referring back to the DPSS designs from Chapter 5, one can now appreciate the necessity for either (a) choosing a crystal that will critically phase match at a temperature where the pump diode will produce an optimal wavelength for pumping or (b) incorporating a separate TEC for the non-linear crystal (or, in most cases, the hybrid vanadate/SHG crystal).

Non-linear materials will phase match depending on the type of material and the cut. As illustrated in Figure 7.4, type I phase matching requires the original waves (of frequency f_1 and f_2) to be in the same polarization—the summed frequency f_3 then appears in a polarization orthogonal to the original. (Extraction of this new wavelength, then, may be accomplished either by filtration or selected by polarization.) Type II phase matching requires that the original components be orthogonal to each other. A particular crystal can work in type I or type II applications depending on the way the crystal is cut and the orientation of the crystal.

The choice of type of crystal usually depends on the polarization of the laser in use. Assuming the laser is polarized, either naturally (for example, where the amplifier material itself is polarized) or through inclusion of an intra-cavity optical element (e.g. a polarizer, often to facilitate the use of an EOM Q-switch), a type I crystal is required since both fundamental photons will be in the same linear polarization. The resulting second-harmonic radiation will be polarized in the opposite plane as the fundamental radiation.

Although it has been assumed until now that f_1 and f_2 are at the same frequency, this is not necessary since harmonic generation in its most generic form is a frequency summing process, and so two different wavelengths may be used. This will be explored further in this chapter when considering the applications of third- and fourth-harmonic generation in which two components of different wavelengths are mixed.

7.3   NON-LINEAR MATERIALS

Ultimately, an ideal non-linear material has a large a_2 coefficient and is naturally birefringent, allowing phase matching. Only a few real materials actually meet these requirements.

Following this discussion of phase matching, refer to the DPSS design from Chapter 5.

TABLE 7.2
Parameters of Some Common Non-Linear Materials

Material	Relative Efficiency	Damage Threshold	Wavelength Range	Notes
KDP	1	5 GW/cm²	200nm–1.5µm	Also used for EOMs
KTP	15	1 GW/cm²	350nm–4.5µm	The most common non-linear material
				Preferred for 1064nm→532nm conversion in Nd:YAG (type II)
				Short wavelength limit limits use for SHG
BBO	6	10 GW/cm²	190nm–3.5µm	Wide wavelength range is useful for SHG, THG, FHG, and even Fifth Harmonic of Nd:YAG
				Can be cut type II for THG, type I for other harmonics
LBO	3	18 GW/cm²	160nm–2.6µm	Preferred for 946nm→473nm conversion in Nd:YAG

One of the first high-efficiency non-linear materials developed for harmonic generation was KDP (potassium dihydrogen phosphate). It is still used for many non-linear applications but newer materials, as outlined in Table 7.2 (which lists only a few commercially available materials), feature higher efficiencies and some have higher damage thresholds. There are many others in addition to those listed, and new ones are constantly under development. Aside from the basic parameters listed, there are other factors to consider when choosing a material, such as the availability of the correct type of material (I or II) at the wavelength required and the ability of the material to phase match a given set of wavelengths since the amount of birefringence varies (i.e. for a given set of wavelengths, the same index of refraction must be available in both axes at some particular angle).

7.4 PRACTICAL CONVERSION EFFICIENCY

Aside from the non-linear coefficient, which dictates the efficiency of a material from a "quantum" perspective, the efficiency of conversion to harmonic radiation, from an implementation perspective, depends on the inverse of the area of the beam—the smaller the area, the higher the intensity and so the larger the non-linear effect that will be observed.

Second-harmonic generation was first observed in quartz, which exhibits very little birefringence. Due to lack of phase matching, efficiency was only 0.000001%.

Conversion efficiency is also proportional to the square of the length of the harmonic generator and the square of incident power as follows:

$$P_{HARMONIC} \propto P_{FUNDAMENTAL}^2 \times l_{CRYSTAL}^2 \times \frac{1}{\pi r^2} \qquad (7.9)$$

Practically, area of the beam is often limited by the application. When used within the cavity of a laser (the most common arrangement since the intra-cavity power is often a hundred times more powerful than the output, assuming a 99% reflecting OC), the beam is the same as the mode within that cavity. Length of the harmonic crystal is limited to some extent as well, by cost and availability.

The largest effect that can be controlled, therefore, is incident power. Doubling the power of this beam quadruples the amount of harmonic radiation produced and so for the ultimate in conversion efficiency it is desirable to operate the harmonic converter at power densities close to the damage threshold for the crystal.

7.5   APPLICATIONS TO LASER DESIGN

In designing a frequency-doubled laser, the first consideration is whether to use the harmonic generator inside or outside the cavity. Intra-cavity harmonic generators are attractive since the intensity inside the cavity is much higher than the power outside the cavity—between ten and one hundred times higher depending on the

---

EXAMPLE 7.1   INTRA- AND EXTRA-CAVITY INTENSITIES

Consider a flashlamp-pumped Q-switched Nd:YAG laser rated at 500mJ at 1064nm with a pulse length of 10ns. Assuming a "square" pulse, the power of the pulse is:

$$P = \frac{E}{t} = \frac{0.5J}{10 \times 10^{-9}}$$

The power of the output beam is hence 50MW. Assuming a 10% OC, the intra-cavity power is 500MW.

And with a 2mm beam diameter—the rod is 6mm in diameter, but the actual output beam is quoted as having a smaller beam diameter and is TEM$_{00}$ mode—the irradiance (called *intensity* by convention throughout most of this text) of the laser output is thus:

$$I = \frac{P}{A} = \frac{50 \times 10^6 W}{\pi \, 0.1cm^2}$$

transmission of the OC—so conversion efficiency of a small DPSS laser may be drastically enhanced by placing the harmonic generator within the cavity. In a Q-switched laser, however, the peak intra-cavity power may be too high and may actually exceed the damage threshold for the non-linear material in use, requiring the use of an external generator.

Obtaining third and fourth harmonics requires either a crystal with a large a_3 (or a_4) coefficient or, owing to the fact there are few materials that have large non-linear coefficients beyond a_2, use of multiple harmonic converters in series. Figure 7.5 illustrates the common method for both third- and fourth-harmonic generation from a YAG laser. The particular laser in this example is a Q-switched Nd:YAG laser employing an EOM Q-switch and as such, incorporates internal polarizers—the output beam is hence linearly polarized. The process begins with generation of the second harmonic at 532nm using a type I crystal with the result being 532nm radiation produced orthogonal to the original 1064nm radiation. Of course, the conversion process is not 100% efficient and so a large quantity of residual 1064nm radiation passes through the second-harmonic generator (SHG) leaving two components: 1064nm and 532nm at orthogonal polarizations. If third-harmonic production is desired, a type II crystal may now be used, which will sum the frequencies of the two orthogonal components producing 355nm radiation. If fourth-harmonic production is desired, use of a type I crystal will double the 532nm radiation into 266nm. This fourth-harmonic radiation will be polarized in the direction as the 1064nm fundamental and orthogonal to the residual 532nm radiation which will pass through.

---

or $1.6*10^9$W/cm^2. For the non-linear generator, it is proposed to use KTP, which has a damage threshold of 1.01GW/cm^2. (The damage threshold of KTP is quoted by various manufacturers as being in the range of 0.5GW/cm^2 to 4.6GW/cm^2, with a conservative value from a reputable manufacturer quoted here.) This material will therefore not work for either use, intra- or extra-cavity. Intra-cavity usage is completely impractical (since the irradiance is ten times higher), but usage outside the cavity would be possible, safely, if either the energy of the laser is limited to 316mJ per pulse (which reduces the irradiance of the output to 1.01GW/cm^2) or the beam diameter is expanded to a little under 3mm before passing through the KTP (which will also serve to reduce the irradiance).

Other possibilities include use of a material with a higher damage threshold such as LBO, which has a damage threshold of 18GW/cm^2.

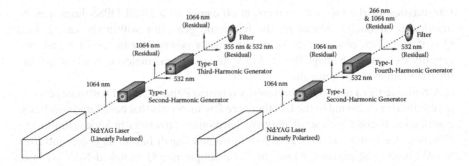

FIGURE 7.5   Third- and fourth-harmonic production schemes.

Of course, production of the fourth harmonic depends on orientation. If the crystal was rotated 90 degrees (and phase-matched), the 1064nm fundamental would be doubled to produce 532nm radiation (already present as residual output passing through the crystal).

7.6   APPLICATION TO DPSS DESIGN

As we saw in Chapter 5, DPSS lasers often incorporate a harmonic generator within the cavity where the intensity of the fundamental is the highest. The non-linear crystal may be a separate crystal or part of a hybrid amplifier/harmonic generator crystal.

A basic design question is, "What is the optimal length of the harmonic generating crystal?" While a long crystal will obviously enhance conversion efficiency, absorption of the fundamental within the laser cavity reduces available intra-cavity power, which ultimately causes a decrease in harmonic output—there is an optimal point at which output is maximized.

Normally, a DPSS (as outlined in Figure 7.6) uses two high reflectors at the fundamental wavelength. Pump radiation passes through the rear reflector (the HR) to end-pump the amplifier and second-harmonic radiation passes through the front reflector (the OC) to become the output beam. Radiation at the fundamental wavelength does not escape the system, keeping the rate of stimulated emission high.

One approach to the problem (although somewhat shortsighted, since an exact solution is considerably more complex[1]) would be to optimize the DPSS laser for maximum intra-cavity 1064nm flux since this should yield the most harmonic output as well. Knowing the basic parameters of the laser one might begin with an optimization similar to those in earlier chapters in which an optimal value for transmission of an OC was computed. The insertion of the harmonic generator crystal represents an additional loss inside the laser cavity in a similar manner to Example 2.3 (in which a distributed loss was considered).

Production of the third harmonic of YAG (at 355nm) illustrates how two different wavelengths can interact in a non-linear material.

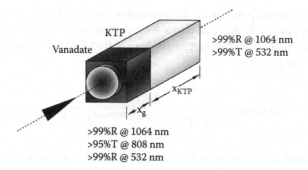

FIGURE 7.6 Optical elements of a DPSS.

Optimizing for intra-cavity power, however, is not enough. On inspection, it can be seen that this will occur when the length of the harmonic generator is zero (i.e. minimal inserted cavity loss), so we must use the expression of Equation (7.9) to compute the expected harmonic output as a function of the length of the generator crystal, squared (we will assume the beam area is constant).

The simple models presented are just that: simple models, based solely on optimization of the intra-cavity fundamental power with the presumption that this will increase harmonic output as well. Analytical solutions exist that are considerably more complex[1] and take various factors into account. The reader is referred to these should a more precise solution be desired.

We now examine two solutions based on two models previously developed in this text.

7.6.1 THE SIMPLE APPROACH

We begin with the simplified expression for output power (Equation 3.31) from Chapter 3:

$$P_{OUT} = \frac{1}{2} P_{SAT} T \left(\frac{2g_0 L}{a+T} - 1 \right) \tag{7.10}$$

In this simplified expression, a represents attenuation in the laser cavity, and T the loss due to the OC. Since we lack an OC at the fundamental wavelength, but now have a second intra-cavity attenuation with the addition of the KTP crystal, we may rewrite this expression for the intra-cavity power as

$$P = \frac{1}{2} P_{SAT} \left(\frac{2g_0 L}{a + a_{KTP}} - 1 \right) \tag{7.11}$$

where a_{KTP} is the round-trip absorption due to the KTP generator crystal in the cavity, which is a function of length as follows:

$$a_{KTP} = 1 - e^{-2\gamma_{KTP} x_{KTP}} \tag{7.12}$$

so that at zero length the absorption of the KTP crystal is zero. The loss due to the crystal is counted twice during a round trip through the laser. Combining the expression for output power with the expression for the efficiency of a harmonic generator (Equation 7.9) and removing a few constant terms to simplify the equation since proportionality is all that is required here to find the peak output, the expected harmonic output power is:

$$P_{HARMONIC} \propto \left(\frac{2g_0 x_g}{2\gamma x_g + (1 - e^{-2\gamma_{KTP}x_{KTP}})} - 1 \right)^2 x_{KTP}^2 \tag{7.13}$$

Analytical solutions are possible but difficult to obtain, and so a numeric solution may be used here in which all known parameters are defined and substitution made for various values of KTP length. Inspection of the output to locate where the peak occurs will reveal the optimal length.

7.6.2 THE RIGROD APPROACH

In a similar approach to that used with the simple expression for output power above, the Rigrod expression from Chapter 4 may be used to predict the harmonic output. In the case of the Rigrod solution presented in Chapter 4, loss due to attenuation must be incorporated as a "point source" loss at a mirror. In the case of this DPSS laser, the intra-cavity loss due to absorption of the harmonic crystal must be treated in the same manner.

We begin with the simplified Rigrod expression for intra-cavity power:

$$I_2 = \frac{I_{SAT}(g_0 x_g + \ln(R_1))}{1 + 1 - R_1 - R_1} \tag{7.14}$$

and define R_1, an average value for mirror reflectivity that incorporates all intra-cavity losses including attenuation in the amplifier and the harmonic crystal, as follows:

$$R_1 = \sqrt{R_{HR}R_{OC}e^{-2\gamma x_a}e^{-2\gamma_{KTP}x_{KTP}}} \tag{7.15}$$

The final expression is then combined with the expression for the efficiency of a harmonic generator (Equation 7.9), a few constant terms are removed for simplicity, and the expected output from the system is found to be:

$$P_{HARMONIC} \propto \left(\frac{g_0 x_g + \ln(R_1)}{2 - 2R_1} \right)^2 x_{KTP}^2 \tag{7.16}$$

In theory, one could differentiate this expression as we did for the Rigrod equation in Chapter 4 to solve for optimal OC transmission (it will yield a mathematically complex equation, although it can be solved). Alternately, a simple numerical solution may be performed to identify the optimal crystal length.

EXAMPLE 7.2   A SMALL GREEN "LASER POINTER" DPSS

Consider a small green "laser pointer" as having the following parameters:

$$R_{HR} = 99.8\%$$
$$R_{OC} = 99.8\%$$
$$g_0 = 25m^{-1}$$
$$\text{Length} = 1mm$$
$$\text{Attenuation} = 2m^{-1}$$

KTP Absorption = $2m^{-1}$ (although some manufacturers claim absorption is "under $1m^{-1}$," the quality of this material varies widely by manufacturer)

From manufacturer specifications, each cavity mirror is 99.8% reflective at the fundamental wavelength of 1064nm. (The HR is highly transmissive at 808nm to allow end-pumping, and the OC is highly transmissive at 532nm to allow the green second-harmonic component to exit the cavity.) The second-harmonic output is then described by Equation (7.13) using the simple model and by Equation (7.16) using the Rigrod model.

A spreadsheet is then set up to generate a table of output power vs. length of the KTP crystal where the crystal varies from 0 to 10mm (the longest available crystal for a small laser such as this). Anchored at the top of the spreadsheet are common parameters including small-signal gain and attenuation of both the amplifier material and the second-harmonic generator material (Figure 7.7).

	A	B	C	D
1	**DPSS Harmonic Crystal Length (Example 7.2)**			
2				
3	Small-signal Gain =	25	m⁻¹	
4	Gain length =	0.001	m	
5	Attenuation =	2	m⁻¹	
6	KTP Attenuation =	2	m⁻¹	
7				
8		Output		
9	KTP length (m)	Simple Model	Rigrod Model	
34	0.0024	4.16468E-05	4.89558E-05	
35	0.0025	4.17376E-05	4.95522E-05	
36	0.0026	4.17535E-05	5.00291E-05	
37	0.0027	4.17003E-05	5.03925E-05	
38	0.0028	4.15838E-05	5.06484E-05	
39	0.0029	4.14089E-05	5.08027E-05	
40	0.003	4.11806E-05	5.08608E-05	
41	0.0031	4.09031E-05	5.08284E-05	
42	0.0032	4.05804E-05	5.07106E-05	

FIGURE 7.7   Spreadsheet parameters.

Column A contains the crystal length in increments of 0.1mm, column B the predicted output using the simple model (Equation 7.13), and column C the predicted output using the Rigrod-based model (Equation 7.16).

The actual spreadsheet formula for output power using the simple model, as found in column B of the spreadsheet, is:

$$B11 = ((((2*\$B\$3*\$B\$4)/$$
$$((2*\$B\$4*\$B\$5)+(1-EXP(-2*\$B\$6*A11))))-1)^2)*A11*A11$$

And for the Rigrod solution,

$$C11 = (((\$B\$3*\$B\$4) + LN((0.992*EXP(-2*A11*\$B\$6))^0.5))/$$
$$(2-2*(0.992*EXP(-2*A11*\$B\$6))^0.5))^2*A11*A11*10$$

where 0.992 represents the fixed loss of the cavity mirrors and the attenuation of the amplifier according to Equation (7.15):

$$R_{HR}R_{OC}e^{-2\gamma x_a}$$

This is then multiplied by the variable loss caused by the KTP (which is a function of KTP crystal length):

$$e^{-2\gamma_{KTP}x_{KTP}}$$

As well, an arbitrary constant of 10 was added to the formulae in column C as a proportionality constant only for display purposes (i.e. to show each peak as approximately the same amplitude on the graph). Neither formulae calculate the actual output power but rather only a proportionality (useful in determining where peak output occurs).

The output from both models is seen, graphed, in Figure 7.8. The two models show good agreement, with the simple model predicting an optimal length

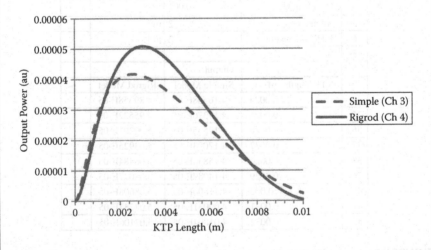

FIGURE 7.8   Intra-cavity power vs. SHG crystal length.

FIGURE 7.9 Hybrid vanadate/SHG crystal (ruler markings in mm).

of 2.6mm and the Rigrod-based model an optimal length of 3mm with the maximum found by inspection of the resulting data. Examination of Figure 7.9, which shows a hybrid crystal of this type (i.e. a 1mm long vanadate crystal), reveals a KTP crystal length very close to that predicted! This design of a DPSS laser incorporating this crystal was discussed in Section 5.9.

REFERENCE

1. Chen, Y.F., et al., 1997. "Single-Mode Oscillation of Compact Fiber-Coupled Laser-Diode-Pumped Nd:YVO/KTP Green Laser." *IEEE Photonics Technology Letters*, Vol. 9, No. 6.

8 Common Lasers and Parameters

We conclude with a quick survey of some common types of lasers including many that have been featured in examples throughout this text.

For each type of laser, parameters are given (a summary of parameters used throughout the book in various examples). These numbers were obtained from research papers or, where possible, from manufacturers (especially for solid-state laser materials) and thus represent values for commercially available materials. Where discrepancies exist between manufacturers' values, either an average or an "accepted" value is used (i.e. one quoted in research literature) as noted.

8.1 CW GAS LASERS

Once the only source of CW laser radiation, gas lasers were discovered shortly after the ruby and have been the dominant source of visible radiation until the past decade, when renewed interest in solid-state lasers has brought us cheaper (and in most cases more robust) replacements. This section provides an introduction to several key types of CW lasers including the HeNe (which is still the de facto standard laboratory laser for many optical testing applications and still has a beam quality rivaling most other types of lasers), the ion laser (once "the only game in town" if you needed visible CW radiation in the green or blue region), and the carbon dioxide laser, which is still an important laser for industrial cutting processes given that these can produce very high output powers and feature a mid-IR wavelength absorbed by many materials.

Provided here is a brief outline of some common gas lasers.

8.1.1 THE HELIUM-NEON (HeNe) GAS LASER

The helium-neon (HeNe) gas laser (introduced in Chapter 1 and used as an example throughout this text) is one of the most basic of all gas lasers and serves as an excellent example in this text. A modern HeNe laser tube, as shown in Figure 8.1, consists of an inner glass bore aligned at the center of a much larger outer tube that serves as a gas envelope. This glass tube is usually packaged inside a cylindrical metal tube for safety and to protect the relatively fragile tube. A discharge is created between the small anode at one end of the tube (right, in the photograph), confined by the small-diameter inner bore where amplification actually occurs, eventually reaching the large cathode region, where the discharge terminates on the large cylindrical aluminum cathode. The cathode and surrounding volume are large to both dissipate heat and to ensure a long life despite the fact that sputtering will bury gas atoms into

FIGURE 8.1   A typical HeNe laser tube.

the cathode and so will lower the internal tube pressure as the laser operates. The large gas ballast volume contains an adequate supply for the life of the tube.

Most HeNe lasers are pumped by an electrical discharge in a low-pressure mixture of helium and neon gases in the approximate ratio of 10:1 and at an approximate total pressure of 2 torr (266 Pa), which allows energetic electrons to flow through the tube. The HeNe laser operates from a moderately high voltage, ranging from about 1kV (for a short tube of under 20cm) to 5kV (for a long tube of almost a meter in length), and a relatively low current of 5mA to 10mA. When these electrons strike a helium atom (which are far more abundant than neon atoms, enhancing the probability that they will indeed strike helium), the light electron transfers energy to the considerably more massive helium atom, raising it to an excited state. Since helium has few energy levels, careful adjustment of the gas pressure of the tube (as a ratio with the applied electric field) will enhance the probability that helium will be excited to a specific energy level 20.61eV above ground. As shown in Figure 8.2, energetic helium atoms then collide with neon atoms to transfer that energy to neon's 20.66eV energy level, which serves as a ULL for all visible transitions in this laser.

Not shown in the figure is the fact that the lower lasing level consists of a series of discrete energy levels, so the HeNe laser can oscillate on nine visible transitions, ranging from 543.5nm in the green to 640.3nm in the red. All visible transitions are CW since the ULL has a much longer lifetime than the LLL, as covered in Section 1.5, where lifetimes of various levels were computed. Several IR transitions are also

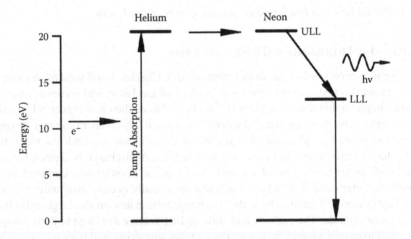

FIGURE 8.2   Simplified energy levels of the HeNe laser.

supported, including a powerful output at 1.15μm (which has a different ULL than shown but shares the same LLL as the visible transitions) and an extraordinarily high-gain transition at 3.39μm (which shares the ULL with the visible transitions), which can often cause design problems for visible HeNe lasers because the gain of this line is so high the laser will preferentially amplify this line, reducing the inversion available for any other line. Cavity optics must suppress this line by either transmitting or absorbing this specific IR wavelength.

From the LLL, excited neon atoms fall to a metastable state (not shown on the figure) at 16.7eV and from there, due to quantum states, cannot lose energy through the emission of a photon to return to ground—a collision with the tube walls is needed to accomplish this (which also dictates that the tube diameter must be small to avoid a buildup of neon atoms in this state).

The vast majority of commercial HeNe lasers are one-piece units with integral cavity mirrors affixed to the ends of the tube. Since the discharge is not particularly energetic, mirrors can be attached directly to the vacuum envelope and windows are not required to protect the delicate optical coatings from attack by exposure to an energetic plasma (as is the case with some lasers we will examine). This arrangement yields a permanently aligned cavity with no maintenance required. Optical cavities are often plano-convex, with one mirror a long-radius concave type. Gain is low, so very high reflectivity optics are required. High-reflectivity dielectric stack mirrors (consisting of alternating quarter-wavelength thick layers of two dielectric materials, often up to 35 such layers) are required, with the HR typically having a reflectivity of over 99.999% and the OC (for the red transition) having a reflectivity of about 99.0% (with the 1% transmitted portion of the intra-cavity beam becoming the output beam)—although the exact value depends on the length of the laser and the wavelength as per Examples 2.7 and 2.8. Other visible transitions have even lower gain and so higher reflectivity optics than 99.0% are required. The OC for a small green HeNe, for which the transition has the lower gain of all visible transitions, is over 99.9% reflecting. Given the quality of the optics required, the cost of these lasers is higher.

Although normally unpolarized, a Brewster plate can be incorporated into the intra-cavity laser optics (usually within the metal stem used to mount one cavity mirror, as per Example 2.9) to polarize the laser. This plate is often made from borosilicate glass, which absorbs the aforementioned 3.39μm wavelength and helps to suppress this transition.

The HeNe laser was used extensively as an example in this text (Table 8.1). See Section 1.5 for specific details on the levels involved (including calculations of lifetimes) and Section 2.6 for an outline of the calculated gain of various other visible lines.

8.1.2 Ion Gas Lasers

The term *ion laser* is reserved for a gas laser in which the active lasing species is ionized, most commonly ionized argon or krypton gas. This definition is worth mentioning since the active lasing species in the Nd:YAG solid-state laser is actually an ion, specifically Nd^{3+}, but this is not called an *ion laser*. At one time, ion lasers were the predominant source of green and blue radiation, but advances in solid-state laser technology have marginalized this laser to limited applications.

TABLE 8.1
HeNe Laser Parameters

Parameter	Value
Lasing wavelength	632.8nm
Cross-section	$3.46 * 10^{-17} \text{m}^2$
ULL lifetime	129ns
Small-signal gain (typical)	0.15m^{-1}
Attenuation	0.005m^{-1}

A typical ion laser consists of a ceramic tube filled with a noble gas (commonly argon or krypton) at low pressure and excited by a high-current discharge, which can range from 10A for the smallest argon-ion tubes to over 70A for a large-frame, water-cooled ion laser. A moderate-sized argon laser with a 5W visible output operates with a current of 40A and a tube voltage of 250V. This represents a total power dissipation of 10kW, so water cooling is required. The smallest ion laser tubes dissipate about 1kW and can be forced-air cooled by a large blower. The high current discharge and the high temperature of the plasma within the tube require the tube to be built in a rather unique way compared to many gas lasers. The tube itself, for instance, is manufactured of ceramic materials to withstand thermal shock during operation, and the discharge is confined to a center bore by tungsten disks within that ceramic tube. A heated cathode is required with such a high-current discharge to enhance the life of the cathode by lowering the power dissipated by this structure. In addition, a moderate to large ion laser tube requires a large magnetic field generated by an electromagnet coaxial to the tube (which also serves as a cooling water jacket) to confine the discharge further to the center of the tube, increasing current density.

Energetic ions in the plasma tube, as well as the presence of UV radiation from a transition that depopulates the LLL, present a hostile environment, precluding the mounting of cavity optics directly onto the tube (at least for all but the smallest ion laser tubes). Most ion laser tubes thus feature Brewster windows at either end and external cavity optics.

As the name implies, energy levels are in an ionic species—argon (Ar^+) or krypton (Kr^+) with a single electron removed, which modifies the resulting energy levels such that visible transitions result. As shown in Figure 8.3, excitation of argon is a two-step process where argon is first ionized and excited to a level 15.76eV above ground. From there, collision with another electron brings the ion to one of many energy levels around 36eV above ground. Excited argon ions can then fall to one of two lower lasing levels producing ten wavelengths in the green, blue, and violet regions of the spectrum. All lower levels have a short lifetime, much shorter than the upper levels, and so CW lasing action is possible on all lines, with this short lifetime brought about by 74nm transition which depopulates these levels.

While lasing is described here in argon, krypton operates in an almost identical manner, producing output in the visible region as well, but with the addition of a yellow line at 568.2nm and a red line at 647.1nm—both important lines and the primary reason for choosing krypton, which has an overall efficiency lower than that

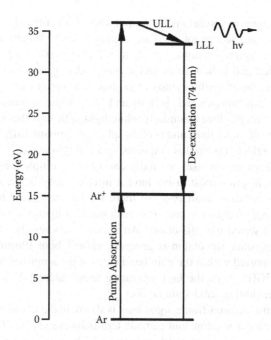

FIGURE 8.3 Simplified energy levels of the argon-ion laser.

of argon. It is also possible for a laser to oscillate on lines that result from transitions in a doubly ionized species (e.g. Ar^{++}). These are all UV lines and were at one time an important source of powerful UV radiation.

Broadband optics allow simultaneous operation of multiple lasing wavelengths, while inclusion of a wavelength selecting optic can restrict operation to a single desired line. In the argon laser, operation on one line does not substantially increase the output power on other lines available since there is little competition between lines. The two most powerful lines are the 488nm blue and 514.5nm green output (which oscillate simultaneously).

The argon laser has higher gain than the HeNe laser, but gain is still low enough that a high-quality cavity is required (although a 5% transmitting OC is common). Lasers often use a long-radius plano-concave cavity in which the OC is concave with a radius much larger than the cavity length. The HR is flat, allowing easy changeover of optics as required (for example, use of a flat HR with a wavelength-specific coating for oscillation on one specific wavelength, a flat HR with a broadband coating for multi-line lasers, or an intra-cavity prism and mirror combination allowing tuning of the laser over any of the available wavelengths).

The argon laser has a moderate gain, the powerful 488nm blue line of the argon-ion laser has a gain of 0.5m⁻¹ (about three times that of the red HeNe transition), and the gain of various krypton-ion laser lines varies, with the important yellow line at 568.2nm having a peak gain of about 0.3m⁻¹ and the 647.1nm red line about 0.1m⁻¹. These are important lines, as they were once the only source of powerful yellow and red radiation. Krypton-ion lasers can operate in multi-line mode producing simultaneous output on several visible output lines. Unfortunately, some lines share a common ULL and so cannot oscillate simultaneously. Where a broad range of lines is desired, krypton and argon gas are mixed in the same ion tube, with krypton producing the red and yellow lines and argon producing a powerful green output that cannot be achieved with krypton alone due to krypton's shared energy levels. (Krypton alone can produce red, yellow, and blue output simultaneously.) Such a mixed-gas laser can produce essentially white light, although the actual output consists of a mixture of many lines and is often split with a prism, diffraction grating, or one wavelength selected from this output using an AOM device.

The output from an ion laser normally consists of multiple longitudinal modes. The long cavity length—most lasers have a mirror cavity between 70cm and 2m in length—also means a small FSR, so these modes are spaced relatively closely. To produce a single-frequency laser (i.e. with single longitudinal mode output), an *etalon* may be inserted into the cavity. An etalon is essentially a "cavity within a cavity"; however, since the etalon is generally short (about 10mm in length), only one mode is supported within the gain bandwidth of the amplifier (which, for an ion laser, is about 5GHz given the high operating temperatures). Etalons can be used with any laser, including solid-state lasers.

The tube from a medium-frame argon laser is shown in the left section of Figure 8.4, in which the Brewster window and cathode terminals are visible. This ceramic tube is inserted into a large magnet, the housing of which also serves as a water-cooling jacket, and into the laser resonator, which holds the cavity optics as well. The complete

FIGURE 8.4   An argon-ion plasma tube (left) and complete laser (right).

TABLE 8.2

Argon-Ion Laser Parameters

Parameter	Value
Lasing wavelength	488.0nm
Cross-section	$2.5*10^{-16}m^2$
ULL lifetime	8ns
Small-signal gain (typical) *	$0.5m^{-1}$
Attenuation	$0.005m^{-1}$

* Gain of this line can vary from $0.45m^{-1}$ to $0.65m^{-1}$ depending on exact tube parameters.

laser, with the tube installed into the magnet housing, is pictured on the right. Like other ion lasers, the tube features a heated cathode (hence the two cathode terminals visible in the photo) as well as an active gas replenishment system that injects gas into the system as required (sputtering occurs during tube operation, which buries gas atoms into the tube and hence reduces pressure—pressure must be kept constant for stability). The complete laser, with a cavity length of about 1m, produces 5W of optical power output and requires 208V three-phase power (at 40A per phase) for a total input power of 14kW. The resulting "wall plug" efficiency is 0.035%. The low efficiency also justifies the need for 9 liters/min of cooling water (see Table 8.2).

8.1.3 THE CARBON DIOXIDE (CO_2) GAS LASER

The carbon dioxide laser is commonly used for industrial processing and cutting applications. The mid-IR wavelength of 10.6μm is absorbed readily by most plastics but is not altogether optimal for cutting many metals. However, the cost per watt rivals that of many other types of lasers.

Like the HeNe laser, the carbon dioxide gas laser uses an electrical discharge for pumping. Consisting of a mixture of carbon dioxide, nitrogen, and helium gases (in increasing proportion), either a high-voltage DC discharge, or in newer lasers, a discharge created by radio-frequency (RF) energy pumps nitrogen to a specific energy level. Then, much like the HeNe laser, energy is transferred from molecular nitrogen to carbon dioxide molecules that then oscillate.

The structure of the laser depends on the pump source. Older designs are primarily of the DC excitation type and are simple structures consisting of a large (generally >10mm) diameter glass tube with a coaxial external water cooling jacket. High-voltage (10kV to 40kV), moderate-current (60mA to 100mA) electrical current through the gas excites the laser. Smaller, compact lasers frequently feature integral optics like a HeNe tube, whereas larger lasers often have tube windows and external optics. The compact laser shown in Figure 8.5 has integral mirrors (the HR on the left, and the OC on the right) as well as a coaxial cooling water jacket. In the case of a larger laser, the length of amplifier tubes is limited, practically, by the power supplies available, and so a folded arrangement is often used (as per Figure 2.5) where

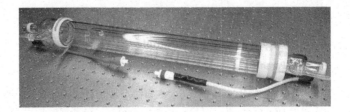

FIGURE 8.5   A DC-excited carbon dioxide laser tube.

several amplifier tubes are used in conjunction with folding mirrors. Some industrial lasers use up to eight tubes of 1.5m in length each to achieve power levels in the multi-kilowatt range. Owing to their simplicity (and the simplicity of the power supply), DC-excited lasers are still quite common.

Newer designs use RF excitation and can have a variety of physical structures ranging from a "traditional" glass tube with external RF electrodes to a waveguide structure consisting of two transverse metal electrodes on the top and bottom of the device and ceramic waveguides on the sides, as diagrammed in Figure 8.6. RF excitation generally leads to a longer tube life than with DC excitation, but at the cost of a more complex power supply and tube design. Other RF-excited designs use a discharge between two large parallel plates and optics (sometimes in the form of an unstable resonator) to produce multiple passes through the discharge in what is called a *slab laser*. RF-excited lasers are often much smaller than DC-excited lasers for the same output power, and designs vary considerably between manufacturers in this competitive market.

Many carbon dioxide lasers are filled at the factory and sealed, so no gas system is required. Larger industrial lasers may feature an external recharge module allowing refilling of the tube(s) as required. Very high power DC-excited lasers may use flowing gas as well, to cool the laser.

Aside from the configurations described here, another popular configuration, used to obtain high pulse energies, is the TEA laser covered in the next section.

As shown in Figure 8.7, all energy levels involved in this laser are vibrational. The pump level, in molecular nitrogen, lies at 0.292eV and represents a vibrational state of the nitrogen molecule. Nitrogen, a simple diatomic molecule with only a few

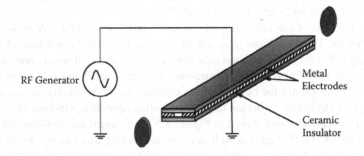

FIGURE 8.6   Structure of a waveguide carbon dioxide laser.

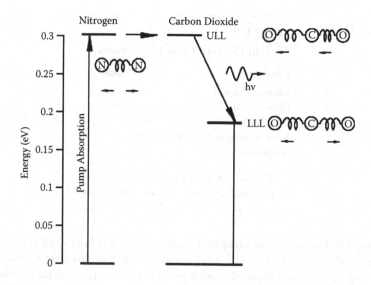

FIGURE 8.7 Simplified energy levels of the carbon dioxide laser.

vibrational modes, has relatively few energy levels, and this is the first oscillatory mode of this molecule. Upon collision with a carbon dioxide molecule, energy is transferred to an asymmetric stretch mode with energy almost identical to that of the pump level (0.291eV). From that state, the energy of the carbon dioxide molecule may drop to one of two lower states producing radiation at 10.6μm or 9.6μm. Only the 10.6μm transition is shown in the figure, in which the LLL is a symmetric stretch mode (the 9.6μm transition terminates in a bending mode). From the lower levels, the molecule must pass through a low-energy bending state before falling to ground.

The carbon dioxide laser oscillates on two series of lines centered at 10.6μm and 9.6μm. Each band originates because the triatomic molecule can possess not only vibrational modes but also rotational modes (where the oxygen atoms rotate around the central carbon). The energy of these modes couples together to form a series of closely spaced discrete energy levels. The large number of lines, collectively, results in the output of the carbon dioxide laser spanning from about 9μm to 11μm. It is possible to "tune" the laser to any of these many lines by using, for example, a gold-coated diffraction grating in place of the HR. This is done when using the laser as a source for IR spectroscopy.

Regardless of the configuration, carbon dioxide lasers possess high gain compared to many other gas lasers and so the quality of the cavity (i.e. reflectivity of optics) need not be as high as many other lasers. In fact, most optics for use in this region of the spectrum are manufactured from metals or metal alloys, either in solid form or deposited as a film onto a substrate, including silicon, gallium-arsenide, gold, copper, and germanium. Glass, for example, is quite opaque to this wavelength and so cannot be used for transmissive optics such as an OC (e.g. where a partially transmissive film is deposited on a transparent glass substrate as is common with many other lasers). In the case of the carbon dioxide laser, one popular material for an OC, zinc-selenide (ZnSe), transmits red light, allowing the use of a red laser for alignment of the cavity.

TABLE 8.3

Carbon Dioxide Laser Parameters

Parameter	Value
Lasing wavelength	10.6μm
Cross-section	$3*10^{-18}m^2$
ULL lifetime	4ms
Small-signal gain (typical) *	$1.0m^{-1}$
Attenuation	$0.05m^{-1}$

* This gain figure is for a sealed, non-flowing gas
laser. Gain increases dramatically with gas flow.

Gain of this laser is large compared to many other gas lasers and is reported in the range of $0.7m^{-1}$ to $1.1m^{-1}$. The small quantum defect of this laser results in high efficiencies, especially compared to other gas lasers and on par with many solid-state lasers. Coupling high efficiencies with a wavelength not achievable by many lasers means this laser will not be obsolete in the near future.

The lifetime of the ULL is extraordinarily long, especially for a gas laser. The situation is complex, though,[1] and mechanisms such as collisions between gas molecules are important here. This is not surprising—most gas lasers have a ULL so short that emission occurs long before collisions of any sort. (See Table 8.3.)

8.2 PULSED GAS LASERS

While essentially any CW gas laser can be operated in pulsed mode, usually by switching the power source on and off rapidly, this section is concerned with lasers that operate "naturally" in pulsed mode. The characteristics of the pulsed lasers outlined here are a pulse shape and power similar to that of a Q-switched laser, which will ablate many materials.

8.2.1 TEA CO_2 LASERS

This version of the carbon dioxide laser utilizes a discharge between two long electrodes that are transverse to the discharge (i.e. they run the length of the "tube"). Such long electrodes, with a small separation between them, allow operation at a relatively high pressure (atmospheric, or even higher). With the increased density of the lasing medium comes an increase in gain, and the output power of such a laser is found to be proportional to the square of pressure: an increase in pressure from that of a "normal" longitudinal carbon dioxide laser (operating at about 60 torr or 8kPa) to atmospheric pressure (760 torr or 101 kPa) results in an increase in output power of over ten times!

The characteristics of the discharge necessary to maintain a voluminous discharge (which fills the entire space between the electrodes) as opposed to a filamentary one (which resembles a localized spark or arc at one or more points on the electrodes) results in a laser that is strictly pulsed and has a very fast rise-time. CW laser action

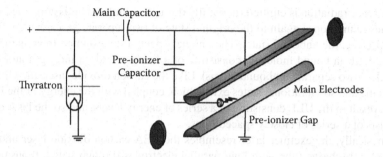

FIGURE 8.8 TEA laser structure and basic circuitry.

is not possible in a laser of this type. In order to form a discharge that fills the entire amplifier the gas volume is first preionized, usually by a spark discharge that emits UV radiation, turning the high-pressure gas into a weakly ionized plasma that readily conducts high currents from a capacitor.

Discharge time is not particularly critical from a quantum viewpoint since the ULL has a very long lifetime, but production of a volume discharge, as required for amplification, requires a very fast rise-time. Electrical components are thus optimized toward this. Electrical connections are made through wide copper strap (with a low intrinsic inductance) rather than simple wires.

Figure 8.8 outlines the structure of a typical TEA laser (the actual design varies by manufacturer), which consists of long transverse electrodes with an optical cavity like that on any other carbon dioxide laser. The main capacitor stores energy, often charging 30kV to 40kV. When the thyratron is fired, this high voltage appears across the main electrodes, but the separation of the plates (coupled with the high gas pressure in the laser) is too large to allow a discharge. Preionizing spark gaps then fire emitting UV radiation, which then weakly ionizes the gas within the laser channel allowing the gas to conduct. The resulting discharge from the main capacitor through the electrodes is voluminous and fills the entire laser channel (which is required for efficient amplification).

The output of the TEA laser is quite powerful. Modest-sized TEA lasers can produce a pulse of a Joule or more with a width of about 100ns resulting in a peak power of 10MW. Pulses of this type have a similar effect upon striking many materials to that of a Q-switched laser, and these lasers often compete with Q-switched YAG lasers for many applications such as marking.

8.2.2 EXCIMER GAS LASERS

An important source of high-power, pulsed UV radiation, excimer lasers are unique in that the ULL and LLL exist as distinct and separate species. An excimer laser uses a gas mixture consisting of a halogen gas (fluorine or chlorine) and an inert gas (argon, krypton, or xenon). When excited with a massive electrical pulse, an EXCIted diMER (or excimer) is formed which defines the ULL. The lifetime of this excited species is very short, though, on the order of 10ns, and so lasing action must ensue quickly to ensure the generation of radiation by stimulated emission from this state. Once radiation is emitted (hopefully through stimulated emission), the excited species returns once again to the original state of two separate gases, which defines the LLL (i.e. the unbound state). The unique aspect of the excimer, then, is the state of the ULL (a bound molecule consisting of a halogen and an inert gas atom) and the LLL (two separate, unbound atoms). Like the carbon dioxide laser, the ULL is a molecule and so vibrational states exist which, coupled with the energy of the bound state, result in the ULL consisting of a series of energy levels, and so the laser output consists of a series of closely spaced lines.

Physically, the excimer laser resembles the TEA carbon dioxide laser and consists of a discharge tube with long parallel electrodes through which thousands of amperes of discharge current flow. A large capacitor stores energy for the laser pulse which is generated when a fast electrical switch (a vacuum tube called a *thyratron*) fires. With an excimer laser mixture, discharge time is more important than with a carbon dioxide laser mixture since the ULL lifetime of the excimer is shorter, dictating a faster discharge (and so, to obtain the fastest discharge times possible, the electrical components of this laser are optimized as such). Another key difference between the excimer and TEA lasers is the highly corrosive mixture used in an excimer, which dictates that this laser must be constructed of materials able to withstand such chemical attack. Excimers used for industrial processes frequently operate at high repetition rates with large input powers requiring a water cooling system.

The gain of an excimer laser is very high so little optical feedback is required. In most cases, an excimer laser will lase with only one (the HR) or no mirrors at all. However, beam quality is drastically improved by the addition of an optical cavity that often consists of an HR (with close to 100% reflectivity) and a low-reflectivity OC (sometimes a quartz blank uncoated on one side to give approximately 4% reflection). Unstable resonators may also be used to produce a specific beam shape.

Commercial excimer lasers operate with one of four common gas mixtures and all produce UV output, as seen in Table 8.4. Gas mixtures shown in the table are typical but vary for each specific laser and manufacturer. Some lasers, for example, use neon as a buffer gas instead of helium, and proportions may vary. The use of a halogen (either fluorine or chlorine) mandates safety requirements that increase the cost of operating this laser (for example, a gas cabinet to contain the halogen gas tank and scrubbers to remove the halogen from waste gas before it is vented). Despite the costs, the excimer remains an important source of intense pulsed UV radiation—the wavelength allowing machining on a finer scale than, say, the TEA carbon dioxide laser.

TABLE 8.4

Common Excimer Gas Mixtures

Laser	Gas Mixture (Typical)	Wavelength (nm)
KrF	0.2% F_2, 5% Kr, Balance He	248
XeF	0.4% F_2, 0.75% Xe, Balance He	350
ArF	0.2% F_2, 14% Ar, Balance He	193
XeCl	0.06% HCl, 1.5% Xe, Balance He	308

8.3 SEMICONDUCTOR (DIODE) LASERS

The most common type of laser, semiconductor diode lasers, are small and inexpensive, lending themselves to a variety of portable applications ranging from communications to media players.

Once restricted to operation in the near-IR region, visible red diodes were developed followed by visible blue and violet diodes. Today it is possible to engineer bandgaps (using novel semiconductor materials) to almost any desired value, allowing semiconductor lasers to be used for many applications once reserved for frequency-doubled solid-state (and before that, gas) lasers.

The actual device is formed where an n-type and p-type semiconductor material are joined. Where the two materials meet, a depletion region is formed that is neutral and devoid of charge carriers since carriers in this region—electrons in the n-type material and holes in the p-type material—will have combined on contact. This is shown in Figure 8.9, where the n-type material is seen to have an excess of electrons in the conduction band and the p-type material an excess of holes in the valence band. Where the two meet a voltage develops (called the *bandgap*) that prevents the flow of charge until an electric field is applied to the junction. As shown in the figure, injection of carriers into the junction, through the application of an electric field, causes electrons on one side of the junction to flow into the depletion region and recombine with carriers on the opposite side of the junction. A minimum voltage (the bandgap) is required to cause mobility of the charge carriers since current will not flow until this voltage is reached, and the resulting energy difference between the two energies (the electrons in the n-type material at a higher energy than the holes in the p-type material) is manifested as the production of a photon with an energy equal to that of the bandgap (and so, by modifying the magnitude of the bandgap, the emission wavelength of the device may be set). This scenario applies to all p-n junctions, even simple diodes (or, for example, a light-emitting diode), but where a laser is concerned an inversion exists between the electrons in the n-type material and the lack of electrons in the valence band near the junction. It is this inversion that gives rise to amplification in this region.

Device structure can assume a variety of forms. The simplest is a homojunction, which consists of a single p- and n-type material. Gain is not well confined

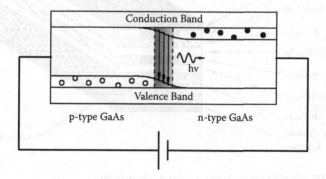

FIGURE 8.9 Charge carriers in a p-n junction.

in this structure—it exists throughout the entire junction, which is a volume of a micrometer or so thick times the top surface area of the device, which is relatively large—so a large amount of current is required to generate an inversion in the entire junction region such that a large amount of heat dissipation results. These devices can hence operate only at low temperatures (e.g. cooled with liquid nitrogen to 77K). A more practical device, and one used in some variation in all common laser diodes, is the heterojunction structure, which consists of several layers of different materials—usually gallium-arsenide (GaAs) and GaAs doped with aluminum, phosphorus, and/or indium. The structure of the device is such that the active gain region is confined in the vertical direction by the interface between the semiconductor materials with many practical devices, using a double heterostructure formed by the use of numerous layers. In the perpendicular direction, the region where gain occurs may be confined by the region where current is present (dictated by the physical layout of the device, which has a stripe electrical contact on the top) or by an index-of-refraction change between the materials on the sides of the device, which forms a waveguide around the active region. Actual structure of devices varies and many structures are possible.

Figure 8.10 shows a simple double heterostructure laser diode, typical of many common laser diodes, in which the gain region is confined on the top and bottom by the interfaces between the p-type GaAs and the p- and n-type AlGaAs, which have different indices of refraction (GaAs has a higher index than AlGaAs). A waveguide structure is hence formed in the vertical direction. The actual order of the layers, as well as the choice of layers shown, depends on the bandgap desired and hence emission wavelength—the active layer may be of different composition and is often doped. As well as the required three layers that form the double heterostructure, the electrical contact is made to p- and n-type GaAs rather than AlGaAs, for better electrical contact.

For a given material, the proportions of impurities determine the bandgap, and by varying these proportions different bandgaps may be engineered. Consider AlGaAs,

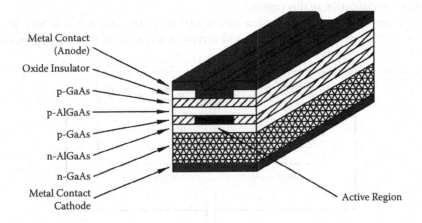

FIGURE 8.10   Double heterostructure laser diode structure.

which is actually GaAs doped with aluminum that replaces some Ga atoms such that the actual formula is $Al_xGa_{1-x}As$. The concentrations can vary, therefore, from $x = 0$ (pure GaAs) to $x = 1$ (pure AlAs) with the corresponding bandgap having emission limits between 870nm and 600nm, respectively. Once these limits are reached, different materials are required for each different wavelength range desired. In the Mid-IR range (2µm to 3µm) GaInAsSb is commonly used, while near-IR diodes (1µm to 1.5µm) often use InGaAsP materials. In the blue and violet regions, the large bandgaps required are developed with GaN and InGaN materials.

In the horizontal direction, the stripe electrode on top of the device defines a narrow region (often only a few micrometers wide) where current density is high enough to allow gain to exceed loss. Not surprisingly, absorption in the semiconductor material (a metal alloy) is large and the material becomes transparent (meaning there is no net absorption) only after a minimum current density is reached. In this device, gain is confined horizontally by current.

The simplest optical cavity consists of simple parallel cleaved ends of the semiconductor which yield a 33% reflection caused by the index-of-refraction change between air ($n = 1.00$) and the GaAs material ($n = 3.7$), as outlined by Equation (2.18). The high gain of semiconductor devices allows use of such low reflectivity optics. This arrangement will yield a laser with two equal output beams—the rear beam usually strikes an integral photodiode used to monitor the actual output of the laser. The resulting device has three terminals: one to power the laser diode, one for photodiode feedback, and the third a common terminal.

A 33% reflectivity for each cavity mirror may not be optimal for all devices and so enhancements may be made by depositing thin-film reflectors directly onto the semiconductor device (especially the HR, since the beam used to monitor the output power of the device does not need to be powerful at all—in fact, too much power will damage the photodiode).

One factor that has inhibited the use of semiconductor lasers for some applications is beam quality, which for a "basic" semiconductor laser is not as high as that of many other lasers. First, the transverse mode of the beam from a semiconductor laser is often elliptical shaped, requiring the addition of focusing optics to reshape the beam (as we have seen in some DPSS designs outlined in Section 5.9). Second, the output of most laser diodes also consists of multiple longitudinal modes which, coupled together, result in a relatively broad emission spectrum. Low coherency restricts usage of these devices in some applications. Because of the short cavity dimensions, the modes of the average laser diode are spaced between 0.1nm and 0.5nm apart and so are easily seen on an OSA, as per Figure 8.11.

Some laser diodes feature wavelength selectors such as distributed Bragg grating (DBG) structures in place of the HR, which are highly wavelength selective and so allow the device to oscillate only on one specific mode. The resulting device is a single longitudinal mode (SLM) device with a very narrow spectral output.

The issue with multiple longitudinal modes can be alleviated with an alternative structure, the Vertical Cavity Surface Emitting Laser (VCSEL), in which the active region is very thin and is sandwiched between two high-reflectivity mirrors (dielectric stacks). The very thin dimensions of the cavity, often a few micrometers, results

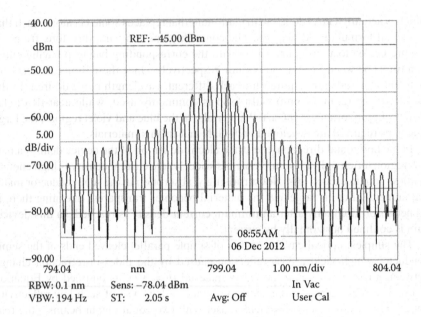

FIGURE 8.11   Longitudinal modes in a laser diode.

in a widely spaced cavity FSR and hence single longitudinal mode operation (since only a single mode falls within the gain profile of the semiconductor material). The other advantage of the VCSEL is the vertical fabrication of the device, which allows manufacturing by similar techniques to those well developed in the semiconductor industry.

Figure 8.12 outlines the basic structure of a VCSEL device that consists of high-reflectivity mirrors (quarter-wave stacks) on the top and bottom of the device and a thin active region between them. The active region often consists of multiple quantum wells as well as insulating layers to restrict current flow to the center of the device and away from the peripheral areas, which are blocked by the top metal electrode regardless.

Although gain of a semiconductor device is high, the thin gain medium of the VCSEL dictates that highly reflective dielectric-stack mirrors be fabricated. The HR, on the bottom of the device, is usually >99.9% reflectivity while the OC is usually only about 1% transmitting. With such a high-quality optical cavity, VCSELs have very low thresholds compared to edge emitting lasers.

As outlined in Chapter 5, diodes require accurate temperature control for wavelength stability. Unlike the ubiquitous HeNe laser, which faithfully oscillates at 632.81646nm and never varies (allowing it to be used as a wavelength standard in the lab), the output of a diode laser is highly dependent on temperature and can easily shift several nanometers as temperature drifts a few degrees. For most applications where wavelength stability is required, diodes are mounted on a Peltier-effect thermoelectric cooler.

The relatively low optical quality (but high efficiency) of laser diode output is one of the primary reasons for the existence of the DPSS laser (covered in the next section).

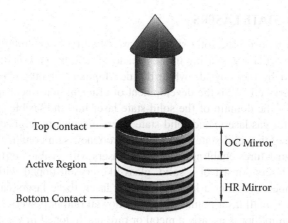

Top Contact

Active Region

Bottom Contact

OC Mirror

HR Mirror

FIGURE 8.12 VCSEL structure.

Few diodes, for example, have a coherent enough beam to allow direct harmonic conversion—the spread in wavelengths present in the laser output preclude efficient phase matching—while the output from a DPSS is, by contrast, quite coherent and the availability of easy intra-cavity access is ideal for harmonic generation. Of course, the new generation of laser diodes operating in the green and other visible regions will change this argument in the coming few years, as will constant improvements to commercial devices. (See Table 8.5.)

An alternative device is the optically pumped semiconductor laser (OPSL), which has a structure much like a VCSEL and is pumped by another diode laser. Often configured much like a thin-disk laser (see Section 8.4.4) in which the semiconductor laser with HR is attached to a heatsink, semiconductor media used in this way exhibit a large pump absorption band, allowing pumping by a wide range of wavelengths. Costs of the system are thus reduced since the pump diode does not need accurate temperature control. Such a laser has all the features of a thin-disk laser, including scalability of output power.

TABLE 8.5
GaAs Semiconductor (Laser Diode) Parameters

Parameter	Value
Lasing wavelength	808nm
Cross-section	$1*10^{-19}m^2$
ULL (recombination) lifetime*	1ns
Small-signal gain (typical)	$10000m^{-1}$
Attenuation	$3500m^{-1}$

* Recombination lifetime is quoted between 1.2ns and 0.6ns with an average value given here.

8.4   SOLID-STATE LASERS

A "technology reborn," the solid-state laser was the first laser. Judging by the number of research articles appearing in the decades that followed this discovery, solid-state lasers fell by the wayside while the development of other lasers, primarily gas lasers, flourished. With the development of efficient non-linear materials came the extension of the domain of the solid-state laser into the visible spectrum (once reserved only for gas lasers) and solid-state lasers once again reigned. The threat to the dominance of the solid-state laser today is from new semiconductor devices with new bandgap structures enabling inexpensive lasers with visible output.

Solid-state lasers consist of a solid, glass-like crystal doped with a lasing ion. Optically pumped by either a lamp or another laser, these lasers can be scaled to anything from a small laser pointer to multi-kilowatt output powers.

The active amplifier is usually a metal or rare-earth doped in varying concentrations into a crystal. The energy levels for the actual laser are provided by the dopant species with the host acting as a carrier, although the host material can affect both the efficiency of the laser and the lasing wavelength in a major way. Common dopants for solid-state lasers include metals such as chromium and titanium and rare-earth metals such as neodymium, ytterbium, and erbium, while hosts can range from simple glass to more exotic oxides and garnets. This material is grown as a crystal that is then machined into a cylindrical rod, rectangle, or large slab to form an amplifier.

Depending on the material, the laser may be three-, four-, or quasi-three-level, and so some solid-state lasers are strictly pulsed while others may operate in CW mode as well. Pumping methods also vary. While all are optically pumped, some require the intense peak powers of a flashlamp and others can be pumped by relatively low-power CW diode lasers. Those pumped by diode lasers can be end-pumped directly through the HR (a logical choice for a compact laser) or from the side in the same manner as a flashlamp-pumped laser.

We will examine each major type of solid-state laser and summarize parameters for these materials (many of which are used in examples throughout the text).

8.4.1   THE RUBY LASER

Ruby, the oldest laser, consists of aluminum oxide (Al_2O_3) doped with chromium ions to approximately 0.05%. The aluminum oxide host is in a crystalline form also known as sapphire which, in pure form, is clear and colorless. Chromium oxide is added as a dopant, with the resulting material being a pink glass-like material, shaped into a cylindrical rod, and optically pumped. The rod is often inserted through the center of a helical flashlamp, with this type of lamp commonly used with ruby given that enormous pump energies are required for this three-level material, which even for a moderate-sized rod can be over 1000 Joules! With a large laser (or one designed for any sort of reasonable repetition rate), these large flashlamps are usually water-cooled, with the lamp and the rod immersed in cooling water.

An outline of the primary elements of a ruby laser can be seen in Figure 8.13, in which the lamp and rod are seen "in the open" and devoid of the normal reflector

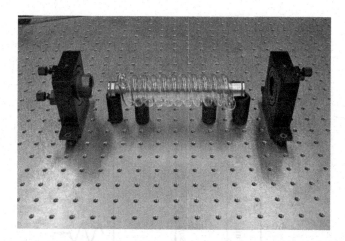

FIGURE 8.13   Components of a ruby laser.

around the pump lamp and water-cooling housing around both that lamp and the ruby rod (both of which usually obscure the view of key elements). For reference, the rod shown (in the center of the flashlamp) is 15cm in length and the lamp normally operates with an input energy of over 3000J!

Ruby oscillates at 694.3nm in the deep red but absorbs pump radiation in two broad bands in the green and violet region of the spectrum, as depicted in Figure 8.14—this is in sharp contrast with the relatively narrow absorption peaks of Nd:YAG. Xenon flashlamps, which output primarily in this region, are used. These lamps emit a somewhat broadband spectrum from the UV to the near IR, but output is somewhat skewed higher in the violet region. Regardless, of all radiation emitted from the flashlamp, only a small percentage is absorbed by ruby.

The energy levels of the ruby laser, as shown in Figure 8.15, represent an almost perfect three-level system. Radiation absorbed at either of two wavelengths (shown

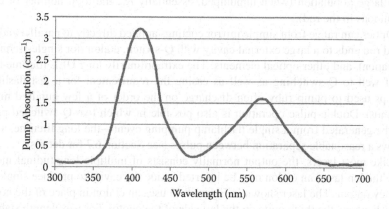

FIGURE 8.14   Pump absorption spectrum of ruby.

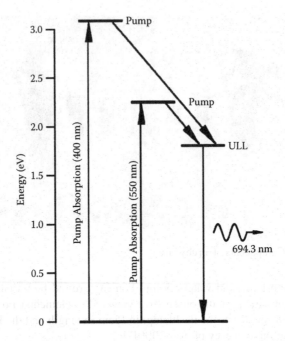

FIGURE 8.15 Simplified energy levels of the ruby laser.

as levels, but in reality these are bands) elevates chromium ions to pump levels where they decay quickly to an ULL with a relatively long lifetime. From there, a transition to ground results in the production of laser radiation.

As a consequence of being a true three-level material, unpumped ruby is a strong absorber of radiation (as illustrated in Example 2.2) at the lasing wavelength. Indeed, immediately after lasing has ensued and the ULL population is lost the material becomes absorbing, since in a three-level laser the LLL is ground and will have a very large population (when unpumped, essentially N_0, the total number of chromium ions in the rod).

Optics can range from simple mirror coatings applied directly to parallel and polished rod ends to a large external cavity with Q-switch, etalon (for single-frequency operation), and other optical elements. The extraordinarily long ULL lifetime lends itself well to Q-switching as well as easing the requirement for a fast flashlamp (lamps used to pump ruby often discharge on the order of a few hundred microseconds). Double-pulse operation is also possible in which two Q-switched pulses can be generated from a single flashlamp pumping event—the long lifetime, again, allows a reasonable separation between pulses (see Section 6.7 for details).

Like most lasers, the output normally consists of multiple longitudinal modes. Like the ion laser, an etalon may be inserted inside the cavity to produce single frequency output. The laser shown in Figure 8.13 uses an etalon in place of the normal OC (shown as the thick optic on the left side of the photo). For wavelength stability, this element is water-cooled. (See Table 8.6.)

TABLE 8.6
Ruby Laser Parameters

Parameter	Value
Lasing wavelength	694.3nm
Cross-section	$2.5*10^{-24}\,m^2$
Doping concentration (0.05% Cr)	$1.58*10^{19}cm^{-3}$
ULL lifetime	3.0ms
Small-signal gain (typical)	$20m^{-1}$

8.4.2   SIDE-PUMPED ND:YAG LASERS

YAG (Yttrium Aluminum Garnet, $Y_3Al_5O_{12}$) is a synthetic crystal commonly used as a host material for many solid-state lasers in which a lasant ion (usually neodymium, but many rare-earth metals may be used) is doped into YAG during manufacture. In a pure form, YAG is colorless until a dopant is added. Neodymium doped YAG (Nd:YAG), for example, is a light violet color depending on the doping concentration.

The most common solid-state laser, the neodymium YAG laser consists of yttrium aluminum garnet doped with neodymium ions at a concentration of about 1%. The shape of the amplifier depends on the size of the laser. In this case we consider high-power YAG lasers in which the amplifier is in the shape of a cylindrical rod and is pumped from the side. Since Nd:YAG is a four-level system, it may be pumped by a CW source (for CW operation) or by a flashlamp for pulsed operation (which, given the high peak powers of a flashlamp, will yield higher gain). Pulsed operation is also possible with a continuous pump source through the use of a Q-switch.

For CW pumping, older lasers used CW krypton arc lamps that resembled flashlamps but operated continuously at currents of 10A to 20A. These lamps, rated for power outputs of many kilowatts, required large amounts of cooling, usually in the form of a closed-loop deionized water system that cools the lamp and exchanges heat with flowing tap water. Newer lasers use arrays of laser diodes, often several located around the rod, to produce pump radiation that is aimed at the laser rod. These arrays are often water-cooled as well, primarily to stabilize the emission wavelength of the diodes to match that of the amplifier material—YAG has a particularly narrow absorption band around 808nm, requiring accurate control of pump wavelength. A typical diode array is seen in Figure 8.16. Radiation from this large diode array (which is water-cooled for wavelength stability) shines into the rectangular slot in the reflector surrounding the YAG rod. There are three such diode arrays surrounding the rod, but only one is visible in the photo. Other components, including the two cavity mirrors, Q-switch, and safety shutter, are external to this pump cavity and are mounted on an optical rail onto which this pump cavity is mounted.

YAG is not naturally birefringent, and so Nd:YAG lasers are unpolarized. Where a polarized YAG laser is desired, intra-cavity elements (such as a polarizing beamsplitter or Brewster plate) are added inside the laser cavity.

FIGURE 8.16   A side-pumped DPSS Nd:YAG laser.

Figure 8.17 shows a simplified view of the energy levels of Nd:YAG. Pumping is usually accomplished around 808nm, where YAG has a strong absorption. Fast decay from this pump level populates a ULL at 0.143eV above ground. Pumping may also be accomplished at 880nm with a lower quantum defect (as covered in Section 5.3). From the ULL, several transitions are possible including 1340nm (not shown), 1064nm, and 946nm. The 1064nm output has the highest gain and hence produces

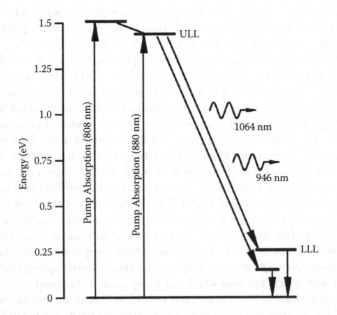

FIGURE 8.17   Simplified energy levels of the Nd:YAG laser.

TABLE 8.7
Nd:YAG Laser Parameters (1064nm)

Parameter	Value
Lasing wavelength	1064nm
Cross-section[*]	$2.8*10^{-19}cm^2$
Doping concentration (approx 1% Nd)	$1.36*10^{20}cm^{-3}$
ULL lifetime	230μs
Small-signal gain (typical, diode-pumped)	$5m^{-1}$
Attenuation (typical)	$0.3m^{-1}$
Pump absorption (808nm)	$3.6cm^{-1}$

[*] While some manufacturers claim cross-section values as large as $6*10^{-23}m^2$, the generally accepted value is given here. The gain figure shown is typical for a CW diode-pumped laser. Flashlamp pumping can yield higher inversions, and higher gain. Note also that the parameters given are for a doping concentration of 1%. Dopant concentration can be as high as 2.5%, at which concentration the effective ULL lifetime decreases and the absorption of pump radiation increases.

the most powerful output (it is by far the most common wavelength for commercial lasers). As a true four-level transition, this transition is relatively insensitive to temperature (i.e. the LLL does not populate appreciably) and so does not require any special cooling to oscillate.

Nd:YAG has several other weaker emission lines with several at longer wavelengths and one at 946nm. This transition terminates on a level only 0.106eV above ground and so operates as a quasi-three-level transition requiring large intra-cavity flux for efficient operation, since a large flux will serve to saturate the re-absorption loss (as outlined in Section 5.5). Because of this limitation, operation at 946nm is often achieved in an end-pumped laser, covered later in this chapter.

Optics for an Nd:YAG laser range from a simple two-mirror cavity to more complex arrangements including Q-switches and modelockers, although the relatively narrow gain bandwidth of the Nd:YAG laser limits the minimum pulse width achievable by this technique. (See Table 8.7.) The technique of Q-switching is covered in Chapter 6.

8.4.3 END-PUMPED ND:YAG LASERS

While side-pumping is used for larger lasers, end-pumped lasers with a short gain medium—in which the amplifier is formed into a short cylinder or a small rectangle (which is more common)—allow compact and simple lasers to be built in a moderate power range. In this arrangement, the amplifier is purposely kept short because of strong absorption of pump radiation.

Nd:YAG, operating at 1064nm, can be end-pumped, but for small lasers vanadate (Nd:YVO$_4$) is often the preferred material. Low threshold pump powers (compared to Nd:YAG) make this material ideal for smaller lasers.

FIGURE 8.18   End-pumping an Nd:YVO4 laser (detail: inset).

Compared to Nd:YAG, vanadate has a higher pump absorption, making use in an end-pumped configuration ideal. The absorption spectrum resembles that of YAG. However, absorption peaks are broader so wavelength control of pump diodes is not as critical. Where a small drift in the wavelength of a pump diode makes a drastic difference in the output of a YAG laser, vanadate is more "forgiving" and a smaller output power swing results, making vanadate ideal for inexpensive lasers operating at room temperature where a temperature swing is expected and inclusion of a thermoelectric cooler impractical (for example, in inexpensive green laser pointers).

Figure 8.18 shows a small DPSS laser in a test jig pumped by an 808nm diode mounted on a TEC. Radiation from the pump diode (in the lower left corner) passes through a beamsplitter to pump a vanadate laser on a separate TEC and mounted on a multi-axis adjustable fixture. The beamsplitter allows monitoring of the pump diode wavelength and power. The inset photo (lower right corner) shows the actual DPSS laser components, which consist of a small piece of vanadate mounted on a copper disk (with a HR deposited directly onto the rear of the crystal) and a separate OC. The output of this DPSS is in the IR at 1064nm (visible as a spot on the IR viewer card in front of the laser). Alternately, a DPSS may be used that incorporates an integral KTP crystal providing green output at 532nm.

Owing to the very high absorption of pump power by vanadate, the material does not lend itself well to side-pumping in the traditional rod configuration. Heavily doped vanadate can have an absorption figure four times that of Nd:YAG. Use of such material in a rod configuration will result in excitation only in an annular ring near the outer periphery of the rod. The center of an evenly pumped 4mm diameter rod would receive only a few percent of the total amount of pump radiation. The resulting mode might resemble a TEM_{01} donut mode, but the excited region is so thin that usage of the amplifier volume and ultimate efficiency would be quite low.

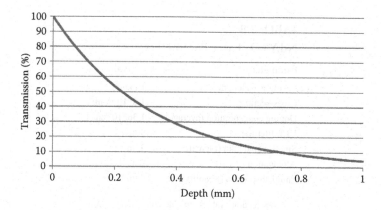

FIGURE 8.19 Absorption of pump radiation in end-pumped vanadate.

Even in an end-pumped configuration, moderate doping levels and a relatively thin amplifier must be employed for efficiency. Figure 8.19 illustrates this by showing the absorption of pump radiation as it travels along a 1mm long, heavily-doped end-pumped vanadate amplifier. At the end of the 1mm gain element, 95% of pump radiation has been absorbed in a single pass, effectively "starving" the end opposite the pump diode.

Unlike YAG, vanadate is naturally birefringent and so the output from a vanadate laser is linearly polarized. Gain is also higher than for Nd:YAG, again a benefit for small lasers and one which allows pumping with small diode lasers.

The shorter ULL lifetime of vanadate means lower stored energy in Q-switched application. This does not preclude Q-switching of a vanadate laser—in fact, the short ULL lifetime lends itself well to fast repetition rate Q-switched lasers—but it does mean lower pulse energies.

Thermal conductivity of vanadate is relatively poor and so operation at high powers exacerbates thermal effects and limits the ultimate power available. Vanadate crystals used for end-pumping are frequently rectangular and usually well under 10mm in length. The material may also be fabricated as small cylindrical rods (usually under 20mm in length). (See Table 8.8.)

While Nd:YAG is not the preferred choice for an end-pumped configuration for operation at 1064nm, end-pumping lends itself well to the use of Nd:YAG on the 946nm transition. This is a quasi-three-level transition, so while the small quantum defect means high potential efficiency, these lasers are subject to high pump thresholds that depend on temperature (as covered in Section 5.4). (See Table 8.9.) Despite potential shortcomings, this transition is of interest, as the frequency-doubled 473nm output serves as a replacement for a blue line of the argon-ion laser. For efficiency, this transition requires operation at large photon fluxes in order to saturate the loss due to re-absorption; hence end-pumping is often used. End-pumping with a small beam radius confines the gain medium within the amplifier, increasing the flux that builds in the laser. To produce the equivalent amount of pumping flux using side-pumping would require an extremely high level of pump power. One limitation of

TABLE 8.8
Nd:YVO₄ Laser Parameters

Parameter	Value
Lasing wavelength	1064nm
Cross-section *	$1.8*10^{-18}cm^2$
Doping concentration (average)	$1.36*10^{20}cm^{-3}$
ULL lifetime	90µs
Small-signal gain (typical)	$25m^{-1}$
Attenuation	$2m^{-1}$
Pump absorption (808nm)	9 to $31cm^{-1}$

* Cross-section values in literature range from $1.14*10^{-18}cm^2$ to $2.5*10^{-18}cm^2$. An average value is given here.

end-pumping, however, is in the ability to obtain a very powerful single-cavity emitter pump diode.

8.4.4 OTHER YAG LASERS

Yb:YAG is an interesting material with an exceedingly small quantum defect. It oscillates at 1030nm and is pumped at 940nm, so very high efficiency is possible. Such as small quantum defect means very little heat is generated inside the material, so thermal effects are reduced when used in a high-power laser. Of course, with such a small quantum defect the material is quasi-three-level and has high pump thresholds. A long ULL lifetime allows Q-switching with significant energy output. (See Table 8.10.)

Unfortunately, as a quasi-three-level material, high intra-cavity intensities are required to saturate the re-absorption loss such that reasonable efficiency is achieved. End-pumping will yield such intensities, but use of a single-emitter device will limit the power of such a design. With a high thermal conductivity, this material is an ideal

TABLE 8.9
Nd:YAG Laser Parameters (946nm)

Parameter	Value
Lasing wavelength	946nm
Cross-section	$5*10^{-20}cm^2$
Doping concentration (approx 1%)	$1.36*10^{20}cm^{-3}$
ULL lifetime	230µs
Small-signal gain (typical, CW pumped)	$5m^{-1}$

Parameters for this material are essentially identical to that employed for 1064nm operation, with the exception of the cross-section, which is, of course, specific to a transition.

TABLE 8.10
Yb:YAG Laser Parameters

Parameter	Value
Lasing wavelength	1030nm
Cross-section	$2.0*10^{-20}cm^2$
Doping concentration (3%)	$4.08*10^{20}cm^{-3}$
ULL lifetime	$960\mu s$
Small-signal gain (typical, CW pumped)[*]	$1.5m^{-1}$
Attenuation	$0.4m^{-1}$

[*] As a relatively new material, gain quoted in various research papers varies from $1.5m^{-1}$ to over $20m^{-1}$. It is dependent on pumping, geometry (e.g. heat removal), and actual doping concentration (which can reach above 30%).

candidate for use with the thin-disk approach (see Figure 8.20 and Section 5.10) to allow high power operation (for which over 1kW of output is possible). As well as use in high-power applications, the second harmonic at 515nm is a potential replacement for the green 514.5nm argon laser.

This same thin-disk arrangement may also be used with semiconductor lasers to form an optically pumped semiconductor laser (OPSL). Unlike solid-state media, semiconductors have a broad absorption band, allowing use with a non-stabilized pump source (i.e. without a TEC). Such technology promises an inexpensive alternative for many lasers.

In addition to YAG and vanadate, other hosts may be used for many solid-state laser ions including sapphire (Al_2O_3), which is the host for chromium in the ruby laser, and glass. Sapphire can, for example, host titanium, which is a broadly tunable medium in the IR and near-IR region (discussed in the next section).

While Nd:YAG is the most common material for a side-pumped arrangement, glass can also be used as a host for neodymium. The resulting Nd:glass material has a high pump threshold but also has a high damage threshold, making it useful

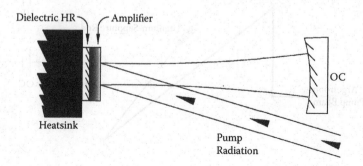

FIGURE 8.20 Thin-disk laser.

for very high power amplifiers. Nd:glass has a lasing wavelength around 1054nm depending on the exact composition of the glass used. Finally, other rare-earth species can also be used with YAG including erbium (which lases at 2940nm), holmium, chromium, and others. Er:YAG, in particular, is often used in the same side-pumped configuration as Nd:YAG.

8.4.5 OTHER SOLID-STATE LASERS

While used most often in a "laser-pumped" configuration, titanium-sapphire (Ti:Saph) was not included under "other YAG lasers" as it is rather unique both in structure and in use. This material is created by doping aluminum oxide (Al_2O_3, the same material used as a host for chromium in the ruby laser) with titanium.

Absorption of Ti:Saph is confined to a narrow region in the green, making pumping from a 514.5nm green argon-ion laser or a 532nm frequency-doubled YAG laser possible. Although flashlamp pumping is possible, the short ULL lifetime makes the pump system on this laser particularly unique, so laser-pumping is far more common, aside from which Ti:Saph is a four-level material and so can operate in CW mode when pumped by a CW source.

The medium supports amplification in a wide range of wavelengths from approximately 690nm in the deep red to 1100nm in the IR. With broadband optics, the laser would oscillate at approximately 800nm where the peak gain occurs. To make the laser tunable, a wavelength selective element must be added to the cavity—most often a birefringent filter, which consists of multiple birefringent plates and is tunable by rotating the filter around the optical axis.

The optical elements of a typical Ti:Saph laser are shown in Figure 8.21. This laser is tunable and is primarily intended for lab applications. Pump radiation from an argon-ion or frequency-doubled YAG laser (shown by the thick line on the diagram) is focused onto a small Ti:Saph rod (usually about 10mm to 15mm in length). The actual cavity is a folded arrangement incorporating a birefringent tuner to select oscillation wavelength. One mirror features a special dielectric coating that allows

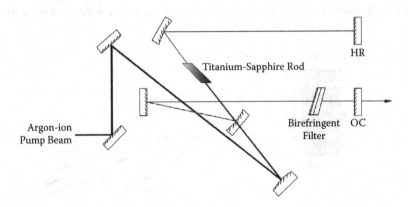

FIGURE 8.21   Optics of a Ti:Saph laser.

pump radiation to pass through it while reflecting infrared and near-IR intra-cavity radiation for the actual Ti:Saph laser cavity. Many other optical arrangements are possible, often incorporating a pump mirror (which allows end-pumping of the rod), as this design does. Some designs use a ring configuration.

In competition with this laser is the OPO (Optical Parametric Oscillator), a tunable oscillator that uses non-linear optical effects. Ti:Saph lasers currently offer higher stability as well as narrow spectral width; however, the necessity for a large pump laser (frequently a large argon-ion laser or frequency-doubled DPSS) makes the OPO an attractive option, especially if a pulsed output is desired for a particular application such as photo-acoustic applications since an OPO is often driven from a Q-switched YAG laser.

As well as producing tunable radiation, Ti:Saph lasers, with an enormous amplification bandwidth, also lend themselves well to the production of ultrashort pulses through modelocking.

REFERENCE

1. Cheo, P.K. 1971. "CO_2 Lasers," in *Lasers,* Eds. A. Levine & A. DeMaria. New York: Marcel Dekker.

Index